Cambridge Tracts in Mathematics
and Mathematical Physics

GENERAL EDITORS
H. BASS, J. F. C. KINGMAN, F. SMITHIES
J. A. TODD and C. T. C. WALL

No. 40

THE LEBESGUE INTEGRAL

THE LEBESGUE INTEGRAL

BY

J. C. BURKILL

Fellow of Peterhouse, Cambridge

CAMBRIDGE UNIVERSITY PRESS

CAMBRIDGE

LONDON · NEW YORK · MELBOURNE

PUBLISHED BY THE PRESS SYNDICATE OF THE UNIVERSITY OF CAMBRIDGE
The Pitt Building, Trumpington Street, Cambridge, United Kingdom

CAMBRIDGE UNIVERSITY PRESS
The Edinburgh Building, Cambridge CB2 2RU, UK
40 West 20th Street, New York NY 10011–4211, USA
477 Williamstown Road, Port Melbourne, VIC 3207, Australia
Ruiz de Alarcón 13, 28014 Madrid, Spain
Dock House, The Waterfront, Cape Town 8001, South Africa

http://www.cambridge.org

First published 1951
Reprinted 1953 1958 1961 1963 1965 1971 1975
First paperback edition 2004

A catalogue record for this book is available from the British Library

ISBN 0 521 04382 4 hardback
ISBN 0 521 60480 X paperback

PREFACE

My aim is to give an account of the theory of integration due to Lebesgue in a form which may appeal to those who have no wish to plumb the depths of the theory of real functions. There is no novelty of treatment in this tract; the presentation is essentially that of Lebesgue himself. The groundwork in analysis and calculus with which the reader is assumed to be acquainted is, roughly, what is in Hardy's *A Course of Pure Mathematics*.

It has long been clear that anyone who uses the integral calculus in the course of his work, whether it be in pure or applied mathematics, should normally interpret integration in the Lebesgue sense. A few simple principles then govern the manipulation of expressions containing integrals.

To appreciate this general remark, the reader is asked to turn to p. 42; calculations such as are contained in Examples 4–8 might confront anyone having to carry through a mathematical argument. Consider in more detail Example 4; the result has the *look* of being right rather than wrong, but the limiting process involved is by no means simple, and the justification of it without an appeal to Lebesgue's principles would be tiresome. Anyone with a grasp of these principles will see that the easily proved fact, that $(1 - t/n)^n$ *increases* to its limit e^{-t}, ensures the validity of the passage to the limit.

The attitude which the working mathematician may take towards the more general concepts of integration has been expressed by Hardy, Littlewood and Pólya in *Inequalities*. After dealing with inequalities between finite sets of numbers and extending them to infinite series, they turn to inequalities between integrals and begin Chapter VI with these preliminary remarks on Lebesgue integrals:

The integrals considered in this chapter are Lebesgue integrals, except in §§ 6·15–6·22, where we are concerned with Stieltjes integrals. It may be convenient that we should state here how much knowledge of the theory we assume. This is for the most part very little, and all that the reader usually needs to know is that there is *some* definition of an integral which possesses the properties specified below. There are naturally many of our theorems which remain significant and true with the older definitions, but the subject becomes *easier*, as well as more comprehensive, if the definitions presupposed have the proper degree of generality.

Since Lebesgue's original exposition a number of different approaches to the theory have been discovered, some of them having attractions of simplicity or generality. It is possible to arrive quickly at the integral without any stress on the idea of measure. I believe, however, that there is an ultimate gain in knowing the outlines of the theory of measure, and I have developed this first in as intuitive a way as possible.

During several years of lecturing on this topic I must have adopted ideas from so many of the books and papers on it that detailed acknowledgement would now be difficult. My greatest debts are to the classical books of de la Vallée Poussin, Carathéodory and Saks, and the straightforward account (having a similar scope to this) given by Titchmarsh in his *Theory of Functions*. I also wish to record that one of my many debts to G. H. Hardy lay in his encouragement to write this tract.

J. C. B.

September, 1949

Reprinting has allowed me to put some details into § 2·2 which had been left to the reader. The first paragraph of § 2·7 mentioning the role of an axiom of choice in the Lebesgue theory has been recast. I might have helped the reader more by discussing this axiom at its first appearance—on p. 3. in enumerating the sets E_m. To do this now would disturb the type too much, and I can help him most by urging him to read an account of the foundations of the subject such as is given in the books specified on p. 87.

There are other less important alterations.

I thank Mr Ingham and Professor Besicovitch for constructive criticism.

J. C. B.

June, 1958

CONTENTS

THE LEBESGUE INTEGRAL

CHAPTER I

SETS OF POINTS

The refinements of the differential and integral calculus, which form the topic of this tract, largely depend on the properties of *sets of points* in one or more dimensions. This chapter contains those properties that will be needed, in so far as they are *descriptive* and not *metrical*. The rules of algebra applied to *sets* hold whether the members of the sets are points or are objects or concepts of any kind. All that we require for a set E to be defined is that we can say of any given object x whether it is or is not a member of E.

1·1. The algebra of sets. Let E be a set,† the members of which may be of any nature. The *sum* of two sets E_1, E_2 is defined to be the set of objects which belong 'either to E_1 or to E_2 (or to both); the sum is written $E_1 + E_2$. By definition $E_2 + E_1$ is the same as $E_1 + E_2$, no question of order being involved. The definition extends to any finite or infinite number of sets, $E_1 + E_2 + \dots$ being the set of objects belonging to at least one E_n. In the definition of an infinite sum there is no appeal to any limiting process.

The *product* $E_1 E_2 \dots$ of any number (finite or infinite) of sets $E_1, E_2, \dots$ is defined to be the set of objects belonging to every one of the sets E_n. The sets E_n may have no members common to all of them, and the product is then the null set—the set which has no members.

† *Class* and *aggregate* are synonymous with *set*; French *ensemble*, German *Menge*.

If every member of E_1 is a member of E_0 we say that E_1 is contained in E_0 and we write $E_1 \subset E_0$ (or $E_0 \supset E_1$). The set of members of E_0 which do not belong to E_1, may be written $E_0 - E_1$ or alternatively as CE_1, the *complement* of E_1. It is easy to see that, the complements being taken with respect to a fixed E_0,

$$C(E_1 + E_2 + \ldots) = CE_1 . CE_2 \ldots,$$
and
$$C(E_1 E_2 \ldots) = CE_1 + CE_2 + \ldots.$$

Limit sets. If $E_1, E_2, \ldots$ is an infinite sequence of sets, the upper limit, $\overline{\lim} E_n$, is defined to be the set of objects which belong to infinitely many of the E_n. The lower limit $\underline{\lim} E_n$ is defined to be the set of objects, each of which belongs to all but a finite number of the E_n. Clearly $\overline{\lim} E_n \supset \underline{\lim} E_n$. If the sets $\overline{\lim} E_n$, $\underline{\lim} E_n$ are the same, we say that the sequence $E_1, E_2, \ldots$ has a limit, $\lim E_n$.

If a sequence of sets is increasing or decreasing, it has a limit: more precisely,

(a) *If $E_n \subset E_{n+1}$, then $\lim E_n = E_1 + E_2 + \ldots$,*

(b) *If $E_n \supset E_{n+1}$, then $\lim E_n = E_1 E_2 \ldots.$*

To prove (a), write $E = E_1 + E_2 + \ldots$ and observe that if x is a member of $\overline{\lim} E_n$ then x is a member of E; hence $\overline{\lim} E_n \subset E$. The result will now follow if we prove that $E \subset \underline{\lim} E_n$. This is true because any member of E is a member of E_n for some n and so (since the sets form an increasing sequence) for all greater n and therefore of $\underline{\lim} E_n$.

A similar proof holds for (b), the product set possibly being null.

More generally, the upper and lower limits of any sequence of sets, not necessarily monotonic (increasing or decreasing), may be expressed in terms of sums and products. The formulae are

$$\overline{\lim} E_n = (E_1 + E_2 + E_3 + \ldots)(E_2 + E_3 + \ldots)(E_3 + \ldots) \ldots,$$
$$\underline{\lim} E_n = E_1 E_2 E_3 \ldots + E_2 E_3 \ldots + E_3 \ldots + \ldots.$$

The proof is left to the reader.

1·2. Infinite sets. Two sets are called *similar* if there is a one-one correspondence between the members of one and the members of the other. Thus two sets with finitely many members are similar if and only if each has the same number of members. The idea of similarity is the foundation of any theory of infinite numbers. We shall give here only those outlines of this topic which are essential for later chapters.

With infinite sets we have a phenomenon which cannot occur with finite sets, namely, that a set can be similar to a part of itself. For instance, the set of positive integers is similar to the set of even integers or to the set of perfect cubes.

Any set which is similar to the set of all positive integers or to a finite sub-set of them is said to be *enumerable*. The one-one correspondence may be displayed by using the positive integers as suffixes, so that the members of any enumerable set may be specified as $x_1, x_2, x_3, \ldots$

It is clear that any sub-set of an enumerable set is enumerable.

The members of an enumerable set of enumerable sets $E_1, E_2, \ldots$ form an enumerable set.

For let the members of E_m be enumerated as $x_{m1}, x_{m2}, x_{m3}, \ldots$ The members of all the sets then form a double array:

$$x_{11} \quad x_{12} \quad x_{13} \quad \cdots$$
$$x_{21} \quad x_{22} \quad x_{23} \quad \cdots$$
$$x_{31} \quad x_{32} \quad x_{33} \quad \cdots$$
$$\cdots \quad \cdots \quad \cdots \quad \cdots$$

This array can be enumerated as a single sequence, for example, by taking terms along the successive diagonals in order

$$x_{11}, x_{12}, x_{21}, x_{13}, x_{22}, x_{31}, \ldots$$

As a particular case of this, the set of positive rational numbers is enumerable, for they are all included in the set

$$\frac{1}{1}, \frac{1}{2}, \frac{2}{1}, \frac{1}{3}, \frac{2}{2}, \frac{3}{1}, \ldots$$

Clearly, the set of all rational numbers (positive, negative, or zero) is enumerable.

A further application of the same argument proves that *the points of the plane of which both co-ordinates are rational form an enumerable set.*

For if $r_1, r_2, \ldots$ are the rationals enumerated, the rational points of the plane can be displayed as

$$(r_1, r_1) \quad (r_1, r_2) \quad (r_1, r_3) \quad \cdots$$
$$(r_2, r_1) \quad (r_2, r_2) \quad (r_2, r_3) \quad \cdots$$
$$\cdots \qquad \cdots \qquad \cdots \qquad \cdots$$

and form an enumerable set of enumerable sets.

The simplest example of a set which is *not* enumerable is the set of all points of an interval. Take the interval $(0, 1)$ and suppose, on the contrary, that all the numbers between 0 and 1 can be enumerated as $x_1, x_2, x_3, \ldots$. Let each x_n be expressed as a decimal

$$x_n = \cdot u_{n1} u_{n2} u_{n3} \cdots,$$

the u's being numbers from 0 to 9. Write down a new number,

$$y = \cdot v_1 v_2 v_3 \cdots,$$

where v_n is determined from u_{nn} by the rule that $v_n = 1$ if $u_{nn} \neq 1$ and $v_n = 2$ if $u_{nn} = 1$. Then y lies between 0 and 1 and is not the same as any x_n, for it differs from x_n in the nth decimal place. This contradicts the hypothesis that the sequence $x_1, x_2, \ldots$ included all the numbers of $(0, 1)$.

1·3. Sets of points. Descriptive properties. Suppose that E is a set of points on a line.

E is *bounded* if all its points are included in some finite interval.

A point P, of abscissa x, is said to be an *interior point* of E if there is a neighbourhood $(x - \delta, x + \delta)$ of P, every point of which belongs to E.

A set E is said to be *open* if every point of it is an interior point. The simplest open set is an interval $a < x < b$ without its end-points.

A point P, of abscissa x (which may or may not be a point of E) is said to be a *limit-point* of E if any neighbourhood $(x - \delta, x + \delta)$, however small δ, contains a point of E other than P. It follows that every neighbourhood of P contains infinitely many points of E. The set of all limit-points of E is called the *derived set* of E and is denoted by E'.

Weierstrass proved that, if E is a bounded set having infinitely many points, then E' contains at least one point.†

A set E is said to be *closed* if $E \supset E'$. (For example, consider the closed interval $a \leqslant x \leqslant b$.)

It will be convenient to reserve the letters O and Q, with suffixes, for open and closed sets respectively. We shall prove first that these two ideas are complementary.

If Q is closed, then CQ is open (the complement naturally must be taken with respect to an *open* interval).

For let P be a point of CQ. Since Q contains all its limit-points, P is not a limit-point of Q. Therefore there is a neighbourhood of P free of points of Q, and so CQ is open.

If O is open, then CO (taken with respect to a closed interval) is closed.

For no point of O is a limit-point of points of CO.

It is to be observed that the set of all points of the line $(-\infty < x < \infty)$ is both open and closed, and so is the null set (the set which contains no points).

If E is any set, E' is closed.

If E' has a limit-point P, let I be a neighbourhood of P. I contains a point of E', say P_1. Let I_1 be a neighbourhood of P_1 contained in I. Then infinitely many points of E lie in I_1 and so in I. Hence P is a limit-point of E and therefore is a point of E'.

The set $\overline{E} = E + E'$ is called the *closure* of E.

† Hardy, *A Course of Pure Mathematics*, ch. 1.

A linear open set is the sum of an enumerable set of open intervals.
Every point of O is contained in an interval I, consisting
entirely of points of O, whose end-points belong to CO. No
two of the intervals I overlap. To prove that they form an
enumerable set, suppose that $P_1, P_2, \ldots$ are the rational points of
the line, enumerated. With each interval I associate the P_n of
smallest suffix contained in it. We thus have a one-one correspondence between the intervals I and a sub-class of the positive
integers, and the theorem is proved.

The sum of (a finite number or) an infinity of open sets is open.
For if P is a point of ΣO_n it is a point of some O_n. It is an
interior point of O_n and *a fortiori* an interior point of ΣO_n.

The result complementary to the last is that the product of
infinitely many closed sets is closed. The product set may be
null. There is one important case in which we can assert that
it is not null, namely, that of a decreasing sequence of bounded
sets.

If $Q_n \subset (a, b)$ and $Q_n \supset Q_{n+1}$, then ΠQ_n is closed and not null.
Let P_n be the left-hand end-point of Q_n. The P_n have a limit-
point P. P is contained in every Q_n and therefore in ΠQ_n.

1·4. Covering theorems.

In proving theorems about real
functions, we often have the situation that every point of a set
E is associated by some property or other with an interval containing it. In general these intervals form a non-enumerable
set and we desire to select an enumerable (or finite) sub-set
which cover every point of E. The reader may well have first
met this problem in a discussion of the properties of a continu-
ous function $f(x)$ in an interval $a \leqslant x \leqslant b$. If, for a given ϵ, we
associate with a value x the greatest interval within which the
function lies between $f(x) \pm \epsilon$, then it can be shown that a finite
number of these intervals will serve to 'cover' (a, b), and this
yields the property of uniform continuity.†

We prove two covering theorems.

† Hardy, *Pure Mathematics*, pp. 196–201.

LINDELÖF'S THEOREM. *To each point x of a set E corresponds an open interval $I(x)$ containing x. Then there is an enumerable set of these intervals covering E.* The rational intervals of the line (i.e. intervals with rational end-points) form an enumerable set. An interval $I(x)$ containing x includes a rational interval containing x. Thus E is covered by an enumerable set of rational intervals, and *a fortiori* by an enumerable set of the $I(x)$.

THE HEINE-BOREL THEOREM. *If the E of Lindelöf's theorem is bounded and closed, it can be covered by a finite number of the associated $I(x)$.* Let $I_1, I_2, \ldots$ be a Lindelöf covering of E. Let E_n be the part of E outside $I_1 + I_2 + \ldots + I_n$. Then E_n is closed. The theorem follows if, for a sufficiently large n, E_n is null. Suppose that no E_n is null. Then, since the E_n form a decreasing sequence of bounded closed sets, there is a point common to all of them. This point is in E and is in no I_n, and we have a contradiction.

1·5. Plane sets.

The reader will satisfy himself that many of the properties which we have for simplicity established for linear sets are true of sets in higher dimensions. Some care is necessary. The natural plane analogue of a linear interval is a rectangle. It is not true that the general open set in the plane can be decomposed into non-overlapping open rectangles in the same way that a linear open set is the sum of intervals. The standard decomposition of a plane open set is into rectangles, not overlapping but adjoining, i.e. having sides and parts of sides in common. For carrying out this process in a systematic way, and for other purposes, the idea of a *network* is useful.

Take a pair of axes and the lines $x = \pm n$, $y = \pm n$ for all integral values. Denote by G_1 the set of all squares of side 1 so formed. Each square is to be regarded as closed, i.e. its sides belong to it. The lines $x = \pm \frac{1}{2}n$, $y = \pm \frac{1}{2}n$ will now form a set of squares of side $\frac{1}{2}$; call this set G_2. Continue this process of bisection; for every integral m we have a set of squares G_m each

8 SETS OF POINTS

of side $1/2^{m-1}$. Call this construction a network and any square appearing in it a *mesh*. The set of meshes is enumerable.

Any point of the plane is then determined by a sequence of meshes $g_1 \supset g_2 \supset \ldots$, where g_m is a mesh of G_m.

We apply this construction to prove the theorem about plane open sets.

A plane open set is the sum of an enumerable set of closed rectangles.

Let O be the set, and P any point of it. Since O is open, a mesh g_m containing P will, if m is sufficiently large, be wholly contained in O. Then O consists of the meshes of G_1 contained in it, the meshes of G_2 contained in O but not in G_1, the meshes of G_3 contained in O but not in G_1 or G_2, and so on. This proves the theorem.

These remarks about sets and networks in a plane can be extended to higher dimensions.

EXAMPLES ON CHAPTER I

(1) If E_n and F_n are two sequences of sets, prove that

$$\underline{\lim} E_n + \underline{\lim} F_n \subset \underline{\lim} (E_n + F_n) \subset \underline{\lim} E_n + \overline{\lim} F_n \subset \overline{\lim} (E_n + F_n)$$

$$= \overline{\lim} E_n + \overline{\lim} F_n.$$

Establish a similar chain of inequalities for products. Deduce that, if $\lim E_n$ and $\lim F_n$ exist, then so do $\lim (E_n + F_n)$ and $\lim (E_n F_n)$.

(2) If, for any choice of a finite number of values $x_1, x_2, \ldots, x_n$, the sum $\sum_{r=1}^{n} f(x_r)$ is bounded, prove that the set of values of x for which $f(x) \neq 0$ is enumerable.

Deduce that the values of x for which a given increasing function is discontinuous form an enumerable set.

(3) Let E be the set of points

$$x = \frac{1}{2^l} + \frac{1}{3^m} + \frac{1}{5^n},$$

where l, m, n have all positive integral values. What are the first, second and third derived sets E', E'', E'''?

(4) If $f(x)$ is a continuous function, and A is any constant, prove that the set $E(f \geqslant A)$ of values of x for which $f(x) \geqslant A$ is closed.

Prove that the same result is true under the more general hypothesis that $f(x)$ is *upper semi-continuous*, i.e. for each ξ,

$$\overline{\lim_{x \to \xi}} f(x) \leqslant f(\xi).$$

(5) A point of E which is not a limit-point is called *isolated*. A set all of whose points are isolated is called an isolated set. Prove that an isolated set is enumerable.

(6) Prove that the set of maxima of a given function $f(x)$ is enumerable.

(7) A set E for which $E = E'$ is called *perfect*. (The simplest perfect set is a closed interval.) Prove that the following construction gives a perfect set (Cantor's ternary set).

From the closed interval $(0, 1)$ remove the middle third, the interval $(\frac{1}{3}, \frac{2}{3})$, taken as open. Remove the (open) middle thirds of each of the two remaining intervals $(0, \frac{1}{3})$ and $(\frac{2}{3}, 1)$. This will leave four intervals. Remove the middle third of each of them. Continue the process indefinitely. The set of points which remain is perfect.

(8) Prove that the perfect set of Ex. 7 is not enumerable.

B

CHAPTER II

MEASURE

2·1. Measure. Following Borel and Lebesgue we shall aim at assigning to a set of points on a line a number called its *measure* which shall generalize the idea of length. The measure of an interval will be its length. For the theory to be satisfactory we shall want, for example, the measure of the sum of two sets without common points to be equal to the sum of the measures of the sets, that is to say, measure is to be an *additive function* of sets of points.

If O is an open set, the additive property requires us to define its measure mO to be the sum of the lengths of its constituent intervals; it is assumed that this sum is convergent (it will certainly be convergent if O is contained in a finite interval).

2·2. Measure of open sets. Let O_1 and O_2 be open sets with no common points. Since the measure of an open set is the sum of the lengths of its intervals we have

$$m(O_1 + O_2) = mO_1 + mO_2.$$

More generally, we shall prove that, if O_1 and O_2 are any two open sets, then

$$m(O_1 + O_2) + m(O_1 O_2) = mO_1 + mO_2.$$

Taking first a special case, if O_1 and O_2 consist of a finite number (n) of intervals, we can prove the result by induction on n.

Generally, let O_1 and O_2 be any open sets; we can suppose their intervals enumerated (p. 6). Let $\epsilon_\nu \to 0$ as $\nu \to \infty$. Take n such that I_n and J_n, the sums of the first n intervals of O_1 and O_2 respectively, satisfy

$$mO_1 - mI_n < \epsilon_\nu \quad \text{and} \quad mO_2 - mJ_n < \epsilon_\nu.$$

Then $I_n J_n \subset O_1 O_2$, and any prescribed interval of $O_1 O_2$ is in $I_n J_n$ if ν is large enough. Hence, as $\nu \to \infty$,
$$m(O_1 O_2) = \lim m(I_n J_n).$$
The inequalities
$$I_n + J_n \subset O_1 + O_2 \subset (I_n + J_n) + (O_1 - I_n) + (O_2 - J_n)$$
show that, as $\nu \to \infty$,
$$m(O_1 + O_2) = \lim m(I_n + J_n).$$
By the special case of the theorem,
$$m(I_n + J_n) + m(I_n J_n) = mI_n + mJ_n.$$
Let $\nu \to \infty$, and we have the theorem for O_1, O_2.

Take now an enumerable infinity of open sets $O_1, O_2, \dots$. If they are disjoint (i.e. no two have common points), then—just as for two sets—
$$m(O_1 + O_2 + \dots) = mO_1 + mO_2 + \dots,$$
if the right-hand side converges. If the sets overlap, the sign $<$ replaces $=$.

This is an occasion to mention infinite measure. If the lengths of the intervals of an open set form a divergent series we could say that its measure is ∞ and could attach extended meanings to some theorems of this chapter. It is generally, however, more convenient to restrict measure to be finite.

2·3. Measure of closed sets.

If Q is a closed set, and O is any open set of finite measure containing Q, then $O - Q$ will be an open set and the additive property requires that

$$mQ = mO - m(O - Q).$$

We take this as the definition of the measure of a closed set. It must be shown that the value of mQ is independent of the particular O chosen. Let O_1, O_2 be two different O's. From § 2·2
$$mO_1 + m(O_2 - Q) = m(O_1 + O_2) + m(O_1 O_2 - Q),$$
and $\qquad mO_2 + m(O_1 - Q) = m(O_1 + O_2) + m(O_1 O_2 - Q).$
By subtraction, $\quad mO_1 - m(O_1 - Q) = mO_2 - m(O_2 - Q)$,
and the definition of mQ is justified.

We now prove the analogue for closed sets of the property
of § 2·2, namely,

$$mQ_1 + mQ_2 = m(Q_1 + Q_2) + m(Q_1 Q_2).$$

Let O be an open set of finite measure containing $Q_1 + Q_2$.
Let $O_1 = CQ_1$, and $O_2 = CQ_2$, the complements being taken with
respect to O. Then $O_1 O_2 = C(Q_1 + Q_2)$ and $O_1 + O_2 = C(Q_1 Q_2)$.
The result of § 2·2 then gives

$$\{mO - mO_1\} + \{mO - mO_2\}$$
$$= \{mO - m(O_1 + O_2)\} + \{mO - m(O_1 O_2)\},$$

and this is what we want.

We shall in future refer to a proof on these lines as a *proof
by complements*.

As a corollary, measure is additive for closed sets. (Take Q_1
and Q_2 to have no common points.)

2·4. Open and closed sets.

(i) (a) *If* $Q \subset O$, *then* $mQ < mO$.   (*O is assumed not null.*)

    (b) *If* $O \subset Q$, *then* $mO \leqslant mQ$.

The fourth line of §2·3 gives (a), since $m(O - Q) > 0$.

To prove (b), observe that, taking complements with respect
to an open set containing Q, we have

$$C(Q - O) = CQ + O,$$

and that CQ and O are open sets with no common points.
From § 2·2,

$$m(CQ) + mO = m\{C(Q - O)\}.$$

From § 2·3 $mQ - mO = m(Q - O)$,

and the right-hand side is greater than or equal to zero.

(ii) (a) *For a given* O *and given* ϵ, *a* Q *can be constructed, contained in* O, *such that* $mO - mQ < \epsilon$.

    (b) *For a given* Q *and given* ϵ, *an* O *can be constructed, con-
taining* Q, *such that* $mO - mQ < \epsilon$.

(a) Let $I_1, I_2, \ldots$ be the intervals of O. Choose n so large that

$$l = mI_1 + mI_2 + \ldots + mI_n > mO - \tfrac{1}{2}\epsilon.$$

From each end of each I_r $(r = 1, 2, \ldots, n)$ cut off an open interval of length $\epsilon mI_r/4l$. This leaves a finite number of closed intervals forming a closed set Q such that

$$mQ = l - \tfrac{1}{2}\epsilon > mO - \epsilon.$$

This proves (a), and (b) follows by complements.

COROLLARY. The measure of an open set is the upper bound of the measures of closed sets contained in it. The measure of a closed set is the lower bound of the measures of open sets containing it.

2·5. Outer and inner measure. Measurable sets.

Given a set E, we define its *outer measure* m^*E to be the lower bound of the measures of open sets containing E. The *inner measure* m_*E is defined to be the upper bound of the measures of closed sets contained in E. If $m^*E = m_*E$, then E is said to be *measurable* and its measure mE is the common value of m^*E and m_*E.

Some results follow at once from the definitions:

(i) *If* $E \subset (a, b)$, *then* $m^*E = b - a - m_*(CE)$. For, if $O \supset E$, then CO, taken with respect to the closed interval (a, b), is a Q contained in CE.

(ii) $m^*E \geqslant m_*E$. For, if $O \supset E \supset Q$, then, by § 2·4 (i) (a), $mO > mQ$.

(iii) *If* $E_1 \supset E_2$, $m^*E_1 \geqslant m^*E_2$ *and* $m_*E_1 \geqslant m_*E_2$.

For any O containing E_1 contains E_2 and so the lower bound of the measures of O's containing E_1 is greater than or equal to the lower bound of the measures of O's containing E_2.

We must assure ourselves that measure, as defined in this section, when applied to open and closed sets is consistent with the measure of such sets, as originally defined. The measure of an O, defined as the sum of its intervals, is equal to m^*O since O itself is an open set containing O. It is equal to m_*O by

§ 2·4(ii)(corollary). Similarly the two definitions of the measure of a closed set are consistent.

2·6. The additive property of measure.

(i) (a) $m^*(E_1 + E_2) + m^*(E_1 E_2) \leqslant m^* E_1 + m^* E_2$,

   (b) $m_*(E_1 + E_2) + m_*(E_1 E_2) \geqslant m_* E_1 + m_* E_2$.

Let O_1, O_2 be open sets containing E_1, E_2, such that

$$mO_1 - m^* E_1 < \epsilon, \quad mO_2 - m^* E_2 < \epsilon.$$

Then $m^*(E_1 + E_2) + m^*(E_1 E_2) \leqslant m(O_1 + O_2) + m(O_1 O_2)$

$$= mO_1 + mO_2 \leqslant m^* E_1 + m^* E_2 + 2\epsilon.$$

Since ϵ is arbitrary, (a) follows; (b) is complementary.

COROLLARY. *If E_1, E_2 are measurable and disjoint, then $E_1 + E_2$ is measurable and $m(E_1 + E_2) = mE_1 + mE_2$.*

This may be extended to a finite number of sets.

An essential feature of the Borel-Lebesgue concept of measure is that it is additive, not merely for a finite number of sets, but for an enumerable infinity. This is a consequence of the following pair of inequalities.

(ii) (a) $m^*(E_1 + E_2 + ...) \leqslant m^* E_1 + m^* E_2 + ...$

   (b) *If no two E's have common points,*

$$m_*(E_1 + E_2 + ...) \geqslant m_* E_1 + m_* E_2 + ...$$

Let O_n be an open set containing E_n such that

$$mO_n - m^* E_n < \epsilon/2^n.$$

Then $m^*(E_1 + E_2 + ...) \leqslant m(O_1 + O_2 + ...)$

$$\leqslant mO_1 + mO_2 + ... \quad \text{from § 2·2}$$

$$\leqslant \epsilon + m^* E_1 + m^* E_2 + ...$$

Since ϵ is arbitrary, (a) is proved. For (b),

$$m_*(E_1 + E_2 + ...) \geqslant m_*(E_1 + E_2 + ... + E_n) \geqslant m_* E_1 + ... + m_* E_n,$$

from the finite case.

Since this is true for all n, the result follows. We have thus proved the fundamental theorem.

If $E_1, E_2, \ldots$ are measurable and disjoint, then $E_1 + E_2 + \ldots$ is measurable and $m(E_1 + E_2 + \ldots) = mE_1 + mE_2 + \ldots$

Note as a special case that any enumerable set of points has measure zero.

2·7. Non-measurable sets. The reader will ask whether sets exist for which the outer and inner measures differ, and, if so, that examples of such sets shall be produced. The answer depends on fundamental questions in the theory of aggregates which we cannot discuss. (See pp. vi, 87.) An 'axiom of choice' admits as being adequately defined an aggregate containing infinitely many members for which no rule of selection can be laid down in advance. The Lebesgue theory repeatedly appeals to the admissibility of an *enumerable* infinity of arbitrary choices (e.g. the choice of the sets O_n in § 2·6 (ii)). To define a non-measurable set we have to assume a stronger axiom of choice admitting as being adequately defined an aggregate depending on a *non-enumerable* infinity of arbitrary choices.

Construction of a non-measurable set. For simplicity of statement we shall define the set on the circumference of a circle; the circle may then be supposed 'unrolled' into an interval and the set becomes a linear set.

Take a circle of diameter 1. Divide the points of the circumference into sets in such a way that, if P is a point of any particular set, then the points distant $1, 2, \ldots$ from P in either direction along the arc belong to the same set. Take one representative of each of these sets, forming an aggregate E; in this step we appeal to the axiom of choice since it is not possible to specify how the representative is to be chosen. Let E_n, E_{-n} be the sets obtained from E by rotations of n in the two directions round the arc. Then it is clear that the sets $E, E_1, E_{-1}, E_2, E_{-2}, \ldots$ are disjoint and together they fill the circumference. Moreover, they are congruent with one another, and if measurable must all have the same measure. This measure cannot be zero, since then the measure of the sum would be zero, whereas it is in fact π. Nor can the measure be positive, for then the measure of the sum would be infinite. Hence E cannot be measurable.

In higher dimensions there are still more curious phenomena. Hausdorff showed that 'a half of a sphere can be congruent with a third of a sphere'. More precisely, the surface S of a sphere can be divided into four sets A, B, C, D, where D is enumerable, and
$$A \equiv B \equiv C, \quad A \equiv B + C,$$
where $\equiv$ denotes congruence by rotation about suitable axes through the centre of the sphere. The proof of this is outside the scope of this tract;† we have said enough to show the interest and difficulty of the problem of measure.

2·8. Further properties of measure. In this section some miscellaneous results are collected.

It is possible to extend to general sets the ideas about *infinite measure* which were put forward for open sets in § 2·2. A set E may be said to be measurable and to have infinite measure if, for every r, the part of it, E_r, contained in the interval $(-r, r)$ is measurable and $mE_r \to \infty$ as $r \to \infty$.

This convention would give a meaning, for example, to the equation expressing the additive property (at the end of § 2·6) when ΣmE_n diverges.

We now pass to an inequality which supplements those of § 2·6 (i).

(i) *If E_1, E_2 are disjoint,*
$$m^*(E_1 + E_2) \geqslant m^*E_1 + m_*E_2 \geqslant m_*(E_1 + E_2).$$

It will suffice if we write out the proof of the former of these inequalities.

If $F \supset G$, we have from § 2·6 (i) (b),
$$m_*F \geqslant m_*G + m_*(F - G).$$

† Hausdorff, *Mengenlehre* (1914), p. 469.

Let $F = CE_1$ (with respect to some containing O) and $G = E_2$. Then, since

$$CE_1 - E_2 = C(E_1 + E_2),$$

we have $\qquad mO - m^*E_1 \geqslant m_*E_2 + mO - m^*(E_1 + E_2),$

giving the result.

(ii) *If M is measurable, then for any E with m^*E finite,*

$$m^*E = m^*(EM) + m^*(E - EM).$$

The inequality § 2·6 (i) (a) gives at once the relation $\leqslant$ between the left-hand and right-hand sides.

Again, from § 2·6 (i) (a),

$$m^*E + m^*M \geqslant m^*(EM) + m^*(E - EM + M),$$

and, from (i) above,

$$m^*(E - EM + M) \geqslant m^*(E - EM) + m_*M.$$

Since $\qquad\qquad\qquad m^*M = m_*M,$

this gives $\qquad m^*E \geqslant m^*(EM) + m^*(E - EM),$

which is an inequality in the reverse direction to the one we have already.

This property of measurable sets has been taken as fundamental by Carathéodory† in his elegant and general exposition of measure.

(iii) *Lebesgue's condition for measurability. A necessary and sufficient condition for E to be measurable is that, given ϵ,*

$$E = \mathscr{E} + e_1 - e_2,$$

where $\mathscr{E}$ is the sum of a finite number of intervals, and $m^*e_1 < \epsilon$, $m^*e_2 < \epsilon$.

To prove necessity, choose O, containing E, such that $mO < mE + \epsilon$.

Take $O - E = e_2$, and take as $\mathscr{E}$ enough intervals of O for $e_1 = O - \mathscr{E}$ to have measure less than ϵ.

† *Vorlesungen über reelle Funktionen.*

To prove sufficiency, we have on the one hand

$$m^*E \leqslant m^*(\mathscr{E}+e_1) \leqslant m\mathscr{E}+\epsilon,$$

and on the other

$$m^*(CE) \leqslant m^*(C\mathscr{E}+e_2) \leqslant m(C\mathscr{E})+\epsilon,$$

and so $m_*E \geqslant m\mathscr{E}-\epsilon.$

Since ϵ is arbitrary, E is measurable.

The gist of this criterion for measurability is that *a measurable set is 'nearly' (in a metrical sense) a finite set of intervals.* (For two other simple guiding principles in the theory of real functions see Examples 3 and 4 at the end of this chapter.)

Other forms of the necessary and sufficient condition are

(1) $E = O-e_1$ where $m^*e_1 < \epsilon$,

or (2) $E = Q+e_2$ where $m^*e_2 < \epsilon$,

or (3) there exist O, Q with $O \supset E \supset Q$ and $mO - mQ < \epsilon$.

(iv) *If $E_1, E_2, \ldots$ are measurable, so are $E_1 + E_2 + \ldots$ and $E_1 E_2 \ldots$.*

(So far only the special case of the sum of *disjoint* sets has been dealt with, in § 2·6 (ii).)

We show that the product $E_1 E_2 \ldots$ is measurable.

By the condition of (iii) in the form (2), given ϵ, we can find closed sets $Q_1, Q_2, \ldots$ such that $E_n = Q_n + e_n$, where $me_n < \epsilon/2^n$.

Then $E_1 E_2 \ldots = Q_1 Q_2 \ldots + e,$

where $e \subset e_1 + e_2 + \ldots,$

and so $m^*e < \epsilon.$

$Q_1 Q_2 \ldots$ is closed (from § 1·3), and so, by the sufficiency of the condition (iii) (2), $E_1 E_2 \ldots$ is measurable.

The result for the sum $E_1 + E_2 + \ldots$ is complementary.

2·9. Sequences of sets.

(i) *Let $E_1 \subset E_2 \subset \ldots$, and let $E = \lim E_n$. Then, if each E_n is measurable, E is measurable and $mE = \lim_{n \to \infty} mE_n$.*

For $E = E_1 + (E_2 - E_1) + (E_3 - E_2) + \ldots,$

and so

$$mE = mE_1 + m(E_2 - E_1) + \ldots = mE_1 + (mE_2 - mE_1) + \ldots,$$

giving the result.

(ii) *Let $E_1 \supset E_2 \supset \ldots$ be a decreasing sequence of measurable sets. Then $m(\lim E_n) = \lim mE_n$.*

The proof is from (i) by complements. (It must be understood in (ii) that the sets E_n are not all of infinite measure.)

We extend (i) and (ii) from limits to upper and lower limits.

(iii) *If the E_n are measurable and are all contained in a set of finite measure, then $\overline{\lim} \, mE_n \leqslant m(\overline{\lim} \, E_n)$.*

Write $F_n = E_n + E_{n+1} + \ldots$. Then F_n is measurable, $F_n \supset F_{n+1}$

and $\overline{\lim} \, E_n = \lim F_n.$

Therefore $m(\overline{\lim} \, E_n) = \lim mF_n.$

But, since F_n contains each of $E_n, E_{n+1}, \ldots,$

$$mF_n \geqslant \overline{\text{bound}} \, (mE_n, mE_{n+1}, \ldots),$$

and so $\lim mF_n \geqslant \overline{\lim} \, mE_n,$

giving the result.

The necessity of the hypothesis that the E_n are contained in a set of finite measure is illustrated by the example E_n equal to the interval $(n, n+1)$, for which the conclusion is false.

(iv) $\underline{\text{Lim}} \, mE_n \geqslant m(\underline{\lim} \, E_n).$

The proof is complementary to that of (iii). Combining (iii) and (iv) we have

(v) *If the E_n are measurable and are all contained in a set of finite measure, and if they have a limit set E, then $\lim mE_n$ exists and equals mE.*

We extend these results to sequences of sets which are not assumed to be measurable. The following lemma will be useful.

(vi) *If E is any set, there are measurable sets F and G such that*

$$F \subset E \subset G$$

and $mF = m_* E, \quad m^* E = mG.$

The sets F and G may be called respectively a *measurable kernel* and a *measurable envelope* of E.

Let ϵ_n be a decreasing sequence tending to zero.

We can choose a sequence O_n of open sets all containing E such that

$$O_n \supset O_{n+1}, \quad mO_n - m^*E < \epsilon_n.$$

Let G be the product set $O_1 O_2 \ldots$. Then G is measurable and contains E. We have

$$mG \geqslant m^*E$$

and

$$mG \leqslant mO_n$$

for every n. Hence

$$mG = m^*E.$$

A complementary construction yields a set F with the properties stated.

(vii) *Let $E_1 \subset E_2 \subset \ldots$, and let $E = \lim E_n$. Then*

$$m^*E = \lim_{n \to \infty} m^*E_n.$$

Since $E_n \subset E$ for every n, $m^*E_n \leqslant m^*E$ and so $\lim m^*E_n \leqslant m^*E$. It remains to prove the reverse inequality.

Let G_n be a measurable envelope of E_n, so that $mG_n = m^*E_n$.

Write

$$H_n = G_n G_{n+1} \ldots$$

then

$$E_n \subset H_n \subset G_n$$

and so

$$m^*E_n = mH_n = mG_n.$$

Now the H_n form an increasing sequence of measurable sets, and if H is their limit, $E \subset H$.

Therefore

$$m^*E \leqslant mH = \lim mH_n \text{ (from (i))} = \lim m^*E_n.$$

This proves (vii). The complementary result is

(viii) *Let $E_1 \supset E_2 \supset \ldots$. Then $m_*(\lim E_n) = \lim m_* E_n$, if the right-hand side is finite.*

2·10. Plane measure. The theory which has been given
in detail for linear sets in §§ 2·1–2·9 can be extended to higher
dimensions. It is sufficient to consider the plane. The foundation stone is the open set, as it is for linear sets. The measure
of a rectangle is its area, and as the (plane) measure of any
linear set must be zero (since it can be enclosed in rectangles of
arbitrarily small area), it is indifferent whether the sides of a
rectangle are regarded as belonging to it or not.

The measure of an open set is then defined to be the sum of
the measures of the meshes of a network of which it is composed (§ 1·5).

To justify this definition we must prove that the measure so
defined is independent of the particular network taken.

Take two networks, which may have different origins and
axes in different directions. Let the open set O be composed
of the meshes $\sum_1^\infty r_i$ of the first network or of the meshes $\sum_1^\infty s_k$ of
the second.

Using the symbol μ for two-dimensional measure, let us
suppose that $\sum_1^\infty \mu r_i > \sum_1^\infty \mu s_k$. Choose n large enough to make

$$\sum_1^n \mu r_i > \sum_1^\infty \mu s_k.$$

Inside each r_i we can place a closed rectangle R_i and about
each s_k we can circumscribe an open rectangle S_k, maintaining
the inequality

$$\sum_1^n \mu R_i > \sum_1^\infty \mu S_k.$$

But the Heine-Borel theorem (§ 1·4) shows that the last inequality is impossible. For every point of the closed set $\sum_1^n R_i$
is a point of O and is interior to some S_k. The set $\sum_1^n R_i$ can
therefore be covered by a finite number of the S_k, and the sum
of the areas of these S_k will be greater than $\sum_1^n \mu R_i$. This

contradicts the preceding inequality, and the definition of the measure of an open set is justified.

The theory of measure of plane sets (or sets in n dimensions) can be developed as in §§ 2·1–2·9.

Ordinate sets. Let E be a linear set, which we shall suppose to lie on the x-axis. Let Ω be the set of ordinates of height h erected on E, i.e. the set of points (x, y) such that x is in E and $0 \leqslant y \leqslant h$. We shall prove that

$$\mu^*\Omega = hm^*E \quad and \quad \mu_*\Omega = hm_*E.$$

In particular, Ω *is measurable if and only if E is, and then*

$$\mu\Omega = hmE.$$

Enclose E in an open set O with $mO < m^*E + \epsilon$.

Let I be a typical interval of O, say $a < x < b$.

Let R be the open rectangle

$$a < x < b, \quad -\eta < y < h + \eta.$$

Then ΣR is an open set enclosing Ω and having measure less than $(h + 2\eta)(m^*E + \epsilon)$. Since ϵ and η are arbitrarily small, it follows that

$$\mu^*\Omega \leqslant hm^*E.$$

We must now prove that $\mu^*\Omega \geqslant hm^*E$. Take an open set containing Ω; express it as the sum of meshes of a network formed by parallels to the axes. Enlarge each mesh slightly into an open rectangle. This construction can be performed so as to give a set of open rectangles ΣS_n, containing Ω, with

$$\Sigma \mu S_n < \mu^*\Omega + \epsilon.$$

Since each ordinate ($x = x_0$, $0 \leqslant y \leqslant h$) is a closed linear set, the Heine-Borel theorem shows that a finite number of the S_n can be selected which cover it. Choose a single open rectangle $T(x_0)$ containing the ordinate and contained in this finite set of S's which cover it. Then $\Sigma T(x_0)$, summed for x_0 in E, is an open set, and its section by the axis $y = 0$ is a linear open set, say ΣI_n. On each I_n erect an open rectangle R_n of height h.

Then $$hm^*E \leqslant hm(\Sigma I_n) = \mu(\Sigma R_n) \leqslant \mu(\Sigma T)$$
$$\leqslant \mu(\Sigma S_n) \leqslant \Sigma \mu S_n < \mu^*\Omega + \epsilon.$$

Since ϵ is arbitrary, this gives the inequality $\mu^*\Omega \geqslant hm^*E$, and so we have $\mu^*\Omega = hm^*E$.

We now give a 'proof by complements' that $\mu_*\Omega = hm_*E$. Enclose E in O, and let Ω_0 be the ordinate set of height h on O. It is easy to see that Ω_0 is measurable and $\mu\Omega_0 = hmO$.

The first part of the theorem, applied to $\Omega_0 - \Omega$ shows that

$$\mu^*(\Omega_0 - \Omega) = hm^*(O - E).$$

Hence, by subtraction,

$$\mu_*\Omega = hm_*E.$$

2·11. Measurability in the sense of Borel.
The approach to the measure of a set by way of outer and inner measure (§ 2·5) was made by Lebesgue in 1902. In 1898 Borel, having defined the measure of open sets and closed sets, had extended the definition to sets obtainable from them by a finite number or an enumerable infinity of operations of addition, subtraction, multiplication and taking a limit. Sets thus definable are said to be measurable (B); they form a sub-class of the class of sets measurable according to Lebesgue's procedure. The advantage of knowing that a set is measurable (B) is that a definite construction for obtaining it from intervals is implied.

Given any measurable set E, there are sets F and G, measurable (B), *such that*

$$F \subset E \subset G \quad \text{and} \quad mF = mE = mG.$$

For the sets constructed in § 2·9 (vi) are measurable (B) and have the properties stated.

2·12. Measurable functions.
Let $f(x)$ be defined in a measurable set E_0. Denote by $E(f > A)$ the set of points of E_0 at which $f(x) > A$, with a similar meaning for $E(f \geqslant A)$, etc.

The function $f(x)$ is said to be *measurable* if, for all values of the constant A, the sets of one of the four families

$$E(f > A), \quad E(f < A), \quad E(f \geqslant A), \quad E(f \leqslant A)$$

are measurable.

We prove that any one of these four conditions implies the other three.

The sets of the first condition are complementary to those of the fourth, and similarly for the second and third conditions.

To prove that the first condition implies the third, we observe that the set $E(f \geqslant A)$ is the product of the sets $E(f > A - \frac{1}{n})$ for $n = 1, 2, ...$, and so, being the product of measurable sets, is measurable. Similar arguments will complete the proof of the equivalence of the four conditions.

The following sequence of results establishes the general principle that elementary operations performed on measurable functions yield measurable functions.

(i) *If f is measurable and c is a constant, then f + c and cf are measurable.*

This follows easily from the definition.

(ii) *If f and g are finite and measurable, the set $E(f > g)$ is measurable.*

If, for a particular x, $f(x) > g(x)$, there is a rational r lying between them. Hence

$$E(f > g) = \sum_r E(f > r) E(g < r)$$

summed over all rationals r, and is the sum of measurable sets.

(iii) *If f and g are finite and measurable, so are f + g and f − g.*

For $E(f + g > A) = E(f > A - g)$. The function $A - g$ is measurable from (i) and the set $E(f > A - g)$ is measurable from (ii).

(iv) *With the hypothesis of (iii), fg is measurable.*

If $A > 0$, $E(f^2 > A) = E(f > \sqrt{A}) + E(f < -\sqrt{A})$, which shows that the square of a measurable function is measurable. A product is reduced to squares by the identity

$$4fg = (f + g)^2 - (f - g)^2.$$

(v) *Let $f_1, f_2, ...$ be a sequence of measurable functions. Then $M(x)$, the upper bound of the values at x of $f_1, f_2, ...$ is measurable. So is the lower bound.*

For
$$E(M > A) = \sum_n E(f_n > A),$$

which is the sum of measurable sets.

(vi) *The limit of a monotonic sequence of measurable functions is measurable.*

For suppose the sequence is increasing ($f_n \leqslant f_{n+1}$). Then the limit is the same as the upper bound, and (v) gives the result.

(vii) *If $f_1, f_2, \ldots$ are measurable, so are the upper and lower limit functions of the sequence.*

Define $M_n(x)$ to be the upper bound of $f_n(x), f_{n+1}(x), \ldots$. Then, by (v), $M_n(x)$ is measurable. Also $M_n \geqslant M_{n+1}$. By (vi), $\lim M_n$ is measurable. But this is the same function as $\overline{\lim} f_n$.

(viii) *A continuous function is measurable.*

If f is continuous, it is easy to see that the set $E(f \geqslant A)$ contains its limit-points, i.e. it is closed and therefore measurable.

EXAMPLES ON CHAPTER II

(1) Prove that Cantor's ternary set (Ch. I, Ex. 7) has measure zero.

(2) Let $f(x, y)$ be a measurable function of x for each y, and continuous in y for each x. Prove that $\overline{\lim}_{y \to a} f(x, y)$ and $\underline{\lim}_{y \to a} f(x, y)$ are measurable functions of x.

(3) Let $f(x)$ be a measurable function in (a, b). Prove that, given ϵ, there is a continuous function $\phi(x)$ such that $|f(x) - \phi(x)| < \epsilon$ except in a set of measure less than ϵ. (In general terms, 'any measurable function is *nearly* a continuous function'.)

(4) *Egoroff's theorem.* Let the sequence of measurable functions $f_n(x)$ tend to the finite limit $f(x)$ in E. Prove that, given δ, we can find a sub-set of E of measure greater than $mE - \delta$ in which the convergence is uniform.

A rough expression of this important theorem is that every convergent sequence of measurable functions is *nearly* uniformly convergent'.

CHAPTER III

THE LEBESGUE INTEGRAL

3·1. The Lebesgue integral. The idea of the definite integral which has come down through the centuries associates $\int_a^b f(x)\,dx$ with the area bounded by the curve $y = f(x)$, the x-axis and the ordinates $x = a$, $x = b$. Having developed in the last chapter the concept of measure of a plane set of points we can, following Lebesgue, present the idea in a refined form.

Let E be a set of points x (which may in a special case be an interval), and $f(x)$ a function, supposed in the first instance to be positive.

Let Ω be the plane set of points (x, y) such that x takes all values in E and $0 \leqslant y \leqslant f(x)$. Ω can be described as the ordinate-set of the function $f(x)$ on E.

If Ω is plane-measurable we shall say that $f(x)$ *has a Lebesgue integral†* in E, written

$$\int_E f(x)\,dx \quad \text{or} \quad (L)\int_E f(x)\,dx,$$

and its value is $\mu\Omega$.

If $f(x)$ is not restricted to be positive, it can be expressed as the difference of two positive functions

$$f(x) = f_+(x) - f_-(x),$$

where $f_+(x)$ is zero for values of x for which $f(x)$ is negative and is elsewhere equal to $f(x)$, and $f_-(x)$ is zero if $f(x)$ is positive and is otherwise equal to $-f(x)$. If Ω_+, Ω_- are the ordinate sets of f_+, f_-, then the integral of f is defined to be $\mu\Omega_+ - \mu\Omega_-$.

† For some years *summable* was commonly used to mean *integrable* (L).

In many cases we shall give proofs of theorems only for positive f, on the understanding that the extension to a general f can be made by the decomposition given in the last paragraph.

The ordinate set Ω of $f(x)$ has been defined by the 'closed' inequality $0 \leqslant y \leqslant f(x)$. We shall show that, so far as measure is concerned, it is indifferent whether the points on the curve $y = f(x)$ are included or not. (This fact will be needed in § 3·9.) Let Ω_0 be the set defined by x in E and $0 \leqslant y < f(x)$.

The transformation $x' = x$, $y' = (1 + \delta)y$ alters the areas of rectangles in the ratio $1 : 1 + \delta$ and hence the outer and inner measures of any plane sets in the same ratio. This transformation turns Ω_0 into an ordinate set of $(1 + \delta)f(x)$. Ω is contained in this and so

$$\mu^*\Omega \leqslant \mu^*\Omega_0\{(1 + \delta)f\} = (1 + \delta)\mu^*\Omega_0(f).$$

Since δ is arbitrary, $\mu^*\Omega \leqslant \mu^*\Omega_0$. But $\Omega_0 \subset \Omega$ and so $\mu^*\Omega = \mu^*\Omega_0$. Similarly $\mu_*\Omega = \mu_*\Omega_0$, and hence if either Ω or Ω_0 is measurable so is the other and $\mu\Omega = \mu\Omega_0$.

3·2. The Riemann integral.
It is assumed that the reader is acquainted with Riemann's idea of the definite integral of a bounded function.† We summarize his definition, in the form given to it by Darboux. Divide the interval (a, b) into a finite number of parts $\delta_1, ..., \delta_n$; let M_r, m_r be the upper and lower bounds of $f(x)$ for values of x in δ_r. Form upper and lower approximative sums

$$S = \sum_1^n M_r \delta_r, \quad s = \sum_1^n m_r \delta_r.$$

Let J be the lower bound of sums S and j the upper bound of sums s for all modes of subdividing (a, b). If $J = j$, this number is the Riemann integral $(R)\displaystyle\int_a^b f(x)\,dx.$

† See, e.g., Whittaker and Watson, *Modern Analysis*, ch. iv; or Hardy, *Pure Mathematics*, ch. vii (for a continuous integrand).

It can be proved that J and j are not merely bounds of sums S and s but are the unique limits of these sums as the length of the greatest sub-interval δ_r tends to zero.

It is easy to think of a function which is integrable (L) in an interval, but not integrable (R). Let E be the set of rational numbers in (a, b). Let $f(x)$ be the *characteristic function* of E, that is to say, $f(x) = 1$ for x in E and $f(x) = 0$ for other values of x. Then, whatever the mode of subdivision of (a, b), every M_r is 1 and every m_r is 0. Hence $J = b - a$ and $j = 0$, and there is no Riemann integral. But since $mE = 0$, it follows from § 2·10 and § 3·1 that $(L)\int_a^b f(x)dx = 0$.

We shall prove the following consistency theorem.

If $(R)\int_a^b f(x)\,dx$ *exists, so does* $(L)\int_a^b f(x)\,dx$ *and they are equal.*

By the Riemann definition, the interval (a, b) can be sub-divided into parts δ_r such that

$$(R)\int_a^b f(x)\,dx - \epsilon < \sum_1^n m_r\delta_r \leqslant \sum_1^n M_r\delta_r < (R)\int_a^b f(x)\,dx + \epsilon.$$

The rectangles of base δ_r and height M_r form a set containing the ordinate set Ω of $f(x)$. Similarly the rectangles of base δ_r and height m_r form a set contained in Ω. Each of these sets of rectangles is plane-measurable, and we have

$$\sum_1^n m_r\delta_r \leqslant \mu_*\Omega \leqslant \mu^*\Omega \leqslant \sum_1^n M_r\delta_r.$$

Since ϵ is arbitrary, Ω is measurable and $\mu\Omega = (R)\int_a^b f(x)\,dx$, which is what we set out to prove.

3·3. The scope of Lebesgue's definition.

We inquire what restrictions are to be laid on the field of integration E and on the integrand $f(x)$ for the Lebesgue integral $\int_E f(x)\,dx$ to have a meaning.

It is clear from § 2·10 that E must be supposed measurable, for if it is not, the integral of a constant would not have a definite value.

If the ordinate-set Ω of a function $f(x)$ is measurable, then $f(x)$ is a measurable function.

Consider the part $S\Omega$ of Ω contained in the strip S of the plane for which $l \leqslant y < l + h$.

Let E_l denote the set of values of x in E for which $f(x) > l$. Then, as h takes a sequence of values decreasing to zero, the sets E_{l+h} form an increasing sequence with E_l as limit-set.

If SE_l and SE_{l+h} are the points of the strip S whose abscissae are respectively in E_l and E_{l+h}, then

$$SE_{l+h} \subset S\Omega \subset SE_l.$$

Remembering that $S\Omega$ is plane-measurable and using the theorem of § 2·10, we have

$$hm^*E_{l+h} \leqslant \mu(S\Omega) \leqslant hm_*E_l$$

and so
$$m^*E_{l+h} \leqslant m_*E_l.$$

Let h tend to zero and we have, from § 2·9 (vii),

$$m^*E_l = \lim_{h \to 0} m^*E_{l+h} \leqslant m_*E_l.$$

Therefore E_l is measurable, and it will follow that $f(x)$ is measurable provided that the set in which $f(x) = 0$ is measurable.

It is to be observed that the set $E(f = 0)$, whether measurable or not, would give an ordinate-set of measure zero. The set $E(f = 0)$ is in fact measurable being the complement of $E(f > 0)$ with regard to E. E was originally assumed measurable and $E(f > 0)$ has been proved measurable.

Hence $f(x)$ is measurable.

The next section will show that the converse of the last theorem is true, though it is clear that, without some restriction on $f(x)$, $\mu\Omega$ may be infinite. This possibility will be ruled out by assuming for the present that $f(x)$ is bounded, and that E has finite measure.

3·4. The integral as the limit of approximative sums.

Suppose that $f(x)$ is measurable and that

$$A \leqslant f(x) \leqslant B.$$

Take a scale of numbers subdividing the range of variation (A, B) of $f(x)$,

$$A = l_1 < l_2 < \ldots < l_n = B.$$

Let $E_r = E(l_r \leqslant f(x) < l_{r+1})$, $e_r = mE_r$ for $r = 1, 2, \ldots, n-1$, and let

$$E_n = E(f(x) = B), \quad e_n = mE_n.$$

Write

$$S = \sum_1^n l_{r+1} e_r, \quad s = \sum_1^n l_r e_r,$$

(where l_{n+1} is taken to be B).

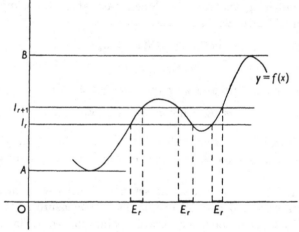

Fig. 1.

We shall prove that:

As the greatest sub-interval $(l_{r+1} - l_r)$ tends to zero, the sums S and s have the common limit $\int_E f(x)\,dx$.

For any scale of subdivision, the numbers S, s satisfy

$$AmE \leqslant s \leqslant S \leqslant BmE.$$

Let J be the lower bound of sums S and j the upper bound of sums s for all modes of subdivision of (A, B).

Let S_1, s_1 be the sums corresponding to any set of numbers dividing (A, B), and S_2, s_2 the sums for another set. Consider now the scale obtained by taking all the numbers of both these sets, and let S', s' be the corresponding sums.

The introduction of a fresh dividing number cannot increase a sum S or decrease a sum s. For suppose a number λ inserted between l_r and l_{r+1}; the increase in S is

$$l_{r+1} mE(\lambda \leqslant f < l_{r+1}) + \lambda mE(l_r \leqslant f < \lambda) - l_{r+1} e_r,$$

which is negative or zero.

Therefore $S_1 \geqslant S' \geqslant s' \geqslant s_2$, that is to say, any upper sum is greater than or equal to any lower sum. Hence $J \geqslant j$.

Moreover, $S - s = \Sigma (l_{r+1} - l_r) e_r \leqslant \epsilon m E$ if $\max (l_{r+1} - l_r) \leqslant \epsilon$.

Therefore $J = j$ and S, s tend to this limit as $\max (l_{r+1} - l_r)$ tends to zero.

We still have to prove that this limit is equal to the Lebesgue integral of $f(x)$ as defined in § 3·1.

Suppose first that $f(x) \geqslant 0$, and let Ω be its ordinate set. Let $\phi(x)$ be the function equal to l_r in $E_r (r = 1, 2, ..., n)$ and $\psi(x)$ the function equal to l_{r+1}, in each E_r; let Ω_ϕ, Ω_ψ be their respective ordinate sets. Then $\Omega_\phi \subset \Omega \subset \Omega_\psi$, and (from § 2·10) $S = \mu\Omega_\psi$, $s = \mu\Omega_\phi$. Hence the common limit of S, s is $\mu\Omega$.

There is no difficulty in extending the proof to cover a bounded function taking both positive and negative values.

We shall now show how the extension to unbounded functions may conveniently be made.

3·5. The integral of an unbounded function.

Suppose that $f(x)$ is measurable and positive. Define an auxiliary function $\{f(x)\}_n$

$$\{f(x)\}_n = f(x) \text{ for } x \text{ such that } f(x) \leqslant n,$$

$$\{f(x)\}_n = n \text{ for } x \text{ such that } f(x) > n.$$

Define $\int_E f(x)\,dx$ as $\lim\limits_{n\to\infty}\int_E \{f(x)\}_n\,dx$ (assumed finite). This agrees

with the definition of $\int_E f(x)\,dx$ as $\mu\Omega$, where Ω is its ordinate-set, for if $(\Omega)_n$ is the ordinate-set of $\{f(x)\}_n$, then $(\Omega)_n$ has limit Ω and $\mu(\Omega)_n$ tends to $\mu\Omega$.

If $f(x)$ is not restricted to be positive, decompose it into $f_+(x)-f_-(x)$ as in § 3·1, and define $\int_E f(x)\,dx$ as being equal to $\int_E f_+(x)\,dx - \int_E f_-(x)\,dx$. This again is in agreement with the definition by ordinate-sets.

If $f(x)$ is integrable in E, so is $|f(x)|$. The converse is true, provided that $f(x)$ is measurable. Also

$$\left|\int_E f(x)\,dx\right| \leqslant \int_E |f(x)|\,dx.$$

If, as above, we write $f(x) = f_+(x)-f_-(x)$, then

$$|f(x)| = f_+(x)+f_-(x).$$

It is easy to see by considering the appropriate ordinate-sets that

$$\int_E |f(x)|\,dx = \int_E f_+(x)\,dx + \int_E f_-(x)\,dx.$$

For the converse, note that, if $f(x)$ is measurable, so are $f_+(x)$ and $f_-(x)$.

Finally, the inequality comes from

$$|a-b| \leqslant a+b,$$

where a and b are the integrals in E of $f_+(x)$ and $f_-(x)$.

It may be said, following the usage for infinite series, that the Lebesgue integral is *absolutely convergent*; that is to say, it integrates only those functions whose modulus is also integrable.

Examples

(1) Verify from the definition that $\int_0^1 x^{-\alpha}\,dx$ exists if $\alpha < 1$.

(2) Let $f(x)$ be positive and bounded in $(\epsilon, 1)$ for each positive ϵ. If $\int_\epsilon^1 f(x)\,dx$ tends to a finite limit l as $\epsilon\to 0$, then $\int_0^1 f(x)\,dx$ exists in the sense of this section and is equal to l.

(3) In Example 2, it is not possible to omit the condition that $f(x)$ is positive. Prove that the function $f(x) = \dfrac{d}{dx}\left(x^2 \sin \dfrac{1}{x^2}\right)$ is not integrable in Lebesgue's sense over $(0, 1)$.

3·6. The integral over an infinite range.

We must review in the light of the Lebesgue theory the interpretation of integrals such as $\int_1^\infty \dfrac{dx}{x^2}$ and $\int_0^\infty \dfrac{\sin x}{x}\,dx$. The simplest way of assigning a meaning to them is as limits, as X tends to infinity, of the integrals $\int_1^X \dfrac{dx}{x^2}$ and $\int_0^X \dfrac{\sin x}{x}\,dx$. It is to be observed that the second of the integrals is not 'absolutely convergent', for it is not difficult to prove that, although $\int_0^X \dfrac{\sin x}{x}\,dx$ tends to a finite limit, $\int_0^X \dfrac{|\sin x|}{x}\,dx$ tends to infinity.

The argument of the first paragraph of § 3·5 will convince the reader that it is indifferent whether an integral of a positive function such as $\int_1^\infty \dfrac{dx}{x^2}$ is regarded as $\lim\limits_{X\to\infty} \int_1^X \dfrac{dx}{x^2}$ or directly as the measure of the appropriate (unbounded) ordinate-set.

If $f(x)$ is not restricted to be positive, then the decomposition

$$f(x) = f_+(x) - f_-(x)$$

of § 3·1 defines its integral over an infinite range (say $-\infty, \infty$) as $\mu\Omega_+ - \mu\Omega_-$ provided that these are finite, in which case $\mu\Omega_+ + \mu\Omega_- = \int |f|$ is also finite, and the integral is absolutely convergent.

The method of the last paragraph would not hold for $\int_0^\infty \dfrac{\sin x}{x}\,dx$, for which the interpretation $\lim\limits_{X\to\infty} \int_0^X \dfrac{\sin x}{x}\,dx$ must be retained.

In future it will be assumed (in the absence of any statement to the contrary) that any range of integration is allowed to be infinite so long as the integral in question is absolutely convergent.

3·7. Simple properties of the integral. This section contains some properties which would be expected of the Lebesgue integral if it is to be a useful concept.

The theorems apply to functions bounded or unbounded, and to ranges of integration finite or infinite. We shall suppose without explicit mention that the functions and sets are measurable. Any proof can be expressed either in terms of ordinate-sets or of approximative sums; we shall choose whichever presentation seems the more direct. The variable of integration will usually for brevity be omitted, and $\int_E f(x)\,dx$ written simply as $\int_E f$.

(i) *If $f \leqslant g$, then $\int_E f \leqslant \int_E g$.*

For, if f is positive, $\Omega(f, E)$, the ordinate-set of f based on E, is contained in $\Omega(g, E)$ and so $\mu\Omega(f) \leqslant \mu\Omega(g)$.

To extend the proof to functions taking both signs, write $f = f_+ - f_-$ and $g = g_+ - g_-$, the suffixes having the meanings assigned to them in § 3·1. Then, in E,

$$f_+ \leqslant g_+ \quad \text{and} \quad f_- \geqslant g_-.$$

Hence

$$\mu\Omega(f_+) \leqslant \mu\Omega(g_+) \quad \text{and} \quad \mu\Omega(f_-) \geqslant \mu\Omega(g_-).$$

Therefore

$$\mu\Omega(f_+) - \mu\Omega(f_-) \leqslant \mu\Omega(g_+) - \mu\Omega(g_-),$$

which is what we have to prove.

COROLLARY. The first mean-value theorem. *If $A \leqslant f \leqslant B$ in a set E of finite measure, then $AmE \leqslant \int_E f \leqslant BmE$.*

(ii) (*The additive property for sets.*) *If E is the sum of a finite number or an enumerable infinity of non-overlapping measurable sets $E_1, E_2, \ldots$, then*

$$\int_E f = \int_{E_1} f + \int_{E_2} f + \ldots,$$

if the left-hand side exists.

If f is positive, let $\Omega = \Omega(f, E)$, $\Omega_n = \Omega(f, E_n)$.

Then $\Omega = \Sigma \Omega_n$, and so $\mu\Omega = \Sigma \mu\Omega_n$.

(iii) *If c is a constant,* $\displaystyle\int_E cf = c\int_E f.$

We give the proof for positive f. Let c be positive. The ordinate-set $\Omega(cf)$ is derived from $\Omega(f)$ by the transformation $x' = x$, $y' = cy$ (homogeneous strain). This alters the areas of rectangles in the ratio $1 : c$, and hence the measures of any sets in the same ratio. This proves the theorem.

If c is negative, say $c = -k$, then $\mu\Omega(kf) = k\mu\Omega(f)$ and

$$\int cf = -\mu\Omega(kf) = -k\mu\Omega(f) = c\int f.$$

(iv) (*The additive property for functions.*) *If f and g are integrable in E, so is $f+g$ and*

$$\int_E (f+g) = \int_E f + \int_E g.$$

We shall suppose first that (i) f and g are bounded, and (ii) E has finite measure, and remove these restrictions in turn.

We start by proving the result when g is a constant k.

Take for f a scale of subdivision, of which l_r is a typical member, giving an approximative sum, $\Sigma l_r e_r = s$ say, for $\displaystyle\int_E f$. The numbers $l_r + k$ then form a scale of subdivision for the function $f + k$, giving an approximative sum, s' say, for $\displaystyle\int_E (f+k)$.

Then $s' = \sum (l_r + k) e_r = s + kmE.$

Taking the limit as $\max(l_{r+1} - l_r)$ tends to zero, we have

$$\int_E (f+k) = \int_E f + \int_E k.$$

Take now g to be any bounded function. Let E_r be the set $E(l_r \leqslant f < l_{r+1})$. Then

$$\int_E (f+g) = \sum_1^n \int_{E_r} (f+g) \quad \text{from (ii)}$$

$$\geqslant \sum_1^n \int_{E_r} (l_r + g) \quad \text{from (i)}$$

$$= s + \int_E g.$$

Similarly, putting l_{r+1} in place of l_r, we have

$$\int_E (f+g) \leqslant S + \int_E g.$$

Taking limits as $\max(l_{r+1} - l_r)$ tends to zero, we deduce that

$$\int_E (f+g) = \int_E f + \int_E g.$$

The result must now be extended to unbounded functions. Suppose first that f and g are both positive. Then, with the meaning assigned to the suffix n in § 3·5,

$$(f+g)_n \leqslant (f)_n + (g)_n \leqslant (f+g)_{2n}.$$

Integrating and letting n tend to infinity, we have the result. If f and g are not always positive, decompose E into measurable sub-sets (six at most) in which f, g, $f+g$ are of constant sign. Consider for example the set in which f and $f+g$ are positive and g is negative. Then $f = (f+g) + (-g)$ is the sum of two positive functions and we have only to apply the last case and transpose the term $\int g$.

If, finally, E has infinite measure, let E_X be the part of E contained in $(-X, X)$. Then, by what we have already proved,

$$\int_{E_X} (f+g) = \int_{E_X} f + \int_{E_X} g.$$

Letting X tend to infinity, we have the result.

Example

If f, g are integrable, establish the integrability of fg under suitable conditions.

3·8. Sets of measure zero.

A property which holds except in a set of measure zero is said to hold *almost everywhere*. The accepted abbreviation for 'almost everywhere' is p.p. (*presque partout*).

(i) *If $f = 0$ p.p. in E, $\displaystyle\int_E f = 0$.*

Let E_1 be the set of measure zero in which $f \neq 0$. Then

$$\int_E f = \int_{E_1} f + \int_{E-E_1} f.$$

The first term on the right-hand side is zero because $mE_1 = 0$ and the second because $f = 0$.

COROLLARY. *If $f = g$ p.p. in E, and one of them is integrable, so is the other and $\displaystyle\int_E f = \int_E g$.*

It is thus possible—and often convenient—to speak of $\displaystyle\int_E f$ although f may be undefined in a subset of measure zero of E.

The next two theorems are in the nature of converses of the last.

(ii) *If $f \geqslant 0$ and $\displaystyle\int_E f = 0$, then $f = 0$ p.p. in E.*

The set $E(f > 0) = \sum_{n=1}^{\infty} E(f > 1/n) = \Sigma E_n$, say. For each n, $mE_n = 0$, for, if not, $\int_E f \, dx \geqslant \int_{E_n} f \, dx \geqslant \frac{1}{n} mE_n > 0$, which contradicts the hypothesis.

Hence $E(f > 0)$ has measure zero, and the theorem is proved.

(iii) *If f is integrable, and* $\int_a^x f = 0$ *for every x in* $a \leqslant x \leqslant b$, *then $f = 0$ p.p. in* (a, b).

If the conclusion is false, then either $f > 0$ in a set of positive measure or $f < 0$ in a set of positive measure—suppose the former. There is then a closed set Q of positive measure contained in the set in which $f > 0$.

Let O be the complement of Q; and ΣI_n its constituent intervals. If I_n is (a_n, b_n), then

$$\int_{I_n} f = \int_a^{b_n} f - \int_a^{a_n} f = 0,$$

by hypothesis.
Hence

$$\int_O f = \Sigma \int_{I_n} f = 0, \quad \text{and} \quad \int_Q f = \int_a^b f - \int_O f = 0.$$

From (ii), $f = 0$ p.p. in Q, which is a contradiction, and the theorem is proved.

3·9. Sequences of integrals of positive functions. The theorems of this section and the next are concerned with the question:

If, as $n \to \infty$, $f_n \to f$, under what conditions can we assert that

$$\int_E f_n \to \int_E f?$$

For Riemann integrals over a finite interval (a, b) the standard sufficient condition is the *uniform convergence* of the sequence f_n to its limit function f.

The conditions for Lebesgue integrals are more general and simpler. The most useful and most easily applied test is that of § 3·10 (D), namely, that f_n is bounded for all n and x in a set E of finite measure.

We begin by observing that the results of § 2·9, (i)–(v), applied to a sequence of ordinate-sets Ω_n and their limit Ω (or more generally their upper and lower limits) yield theorems about the integration of a sequence of positive functions.

Some lemmas are required.

If $f_n \geqslant 0$ and Ω_n is an ordinate-set of f_n, then $\Sigma\Omega_n$ and $\Pi\Omega_n$ are respectively ordinate-sets of $M(x)$ and $m(x)$, the upper and lower bounds at x of $f_1(x), f_2(x), \ldots.$

If $0 \leqslant f_n \leqslant f_{n+1}$, and $f = \lim f_n$, then $\lim \Omega_n$ is an ordinate-set of f.

For any sequence of positive functions f_n, $\overline{\lim} \, \Omega_n$ and $\underline{\lim} \, \Omega_n$ are ordinate-sets of $\overline{\lim} \, f_n$ and $\underline{\lim} \, f_n$.

The assertion about $M(x)$ is easily seen to be true if the ordinate-set Ω_n is taken to be 'open at the top', i.e. the set $(x$ in E, $0 \leqslant y < f_n(x))$ and the ordinate set of $M(x)$ defined similarly by the relations $(x$ in E, $0 \leqslant y < M(x))$.

For $m(x)$ take 'closed' ordinate-sets with defining inequalities $0 \leqslant y \leqslant f_n(x)$ and $0 \leqslant y \leqslant m(x)$.

The second and third paragraphs follow from the first in the same way that (vi) and (vii) of § 2·12 follow from (v).

Using these lemmas, we deduce the following results from § 2·9, (i)–(v).

Suppose that $f_n \geqslant 0$ in the range of integration.

(i) *If $f_n \leqslant f_{n+1}$, and $f = \lim f_n$, then*

$$\int f = \lim \int f_n.$$

(ii) *If $f_n \geqslant f_{n+1}$, and $f = \lim f_n$, then*

$$\int f = \lim \int f_n.$$

(iii) *If, for all n, $f_n(x) \leqslant \psi(x)$, where ψ is integrable, then*

$$\overline{\lim} \int f_n \leqslant \int (\overline{\lim} f_n).$$

(iv) *For any sequence of positive functions f_n,*

$$\int (\underline{\lim} f_n) \leqslant \underline{\lim} \int f_n.$$

(v) *It follows from* (iii) *and* (iv) *that, if* $0 \leqslant f_n \leqslant \psi$, *where ψ is integrable, and f_n tends to a limit f, then*

$$\lim \int f_n = \int f.$$

From the principle of § 3·8, *this still holds if it is only true p.p. that f_n tends to f.*

These statements (i)–(v) are for the most part steps towards the more general results for functions of arbitrary sign given in the next section. The one which has the greatest independent interest is (iv), and we restate it in its most useful form.

FATOU'S LEMMA. *If $f_n \geqslant 0$, and $f_n \to f$ p.p., then*

$$\int f \leqslant \underline{\lim} \int f_n.$$

3·10 Sequences of integrals (integration term by term).
We now suppose that the sign of f_n is unrestricted and give the most useful results.

(*A*) BEPPO LEVI'S THEOREM. *If $f_n \leqslant f_{n+1}$, and $f = \lim f_n$, then*

$$\int f = \lim \int f_n.$$

The theorem (i) of § 3·9 applied to the sequence $f_n - f_1$ gives

$$\int f - \int f_1 = \lim \int f_n - \int f_1,$$

and hence the result.

Notice that if $\int f_n \to \infty$, the theorem admits of the interpretation that then $\mu\Omega$ is infinite, Ω being the ordinate-set of f.

This theorem may be restated in the language of series instead of sequences.

(B) *If* $\Sigma u_n(x)$ *is a series of positive terms, then*

$$\int \Sigma u_n = \Sigma \int u_n,$$

provided that either side is finite.

(C) THEOREM OF DOMINATED CONVERGENCE. *If, for all n,* $|f_n(x)| \leqslant \psi(x)$, *where* ψ *is integrable, and* $f_n \to f$ *p.p., then*

$$\int f = \lim \int f_n.$$

Take the sequence $f_n + \psi$, which is positive and is 'dominated' by the integrable function 2ψ.

Then § 3·9 (iii) gives

$$\overline{\lim} \int f_n + \int \psi \leqslant \int f + \int \psi.$$

And § 3·9 (iv) gives

$$\int f + \int \psi \leqslant \underline{\lim} \int f_n + \int \psi.$$

This pair of inequalities proves the theorem.

The following special case is constantly used.

(D) THEOREM OF BOUNDED CONVERGENCE. *If, in a set E of finite measure,*

$$|f_n(x)| \leqslant M \text{ for all } n, x,$$

and $f_n \to f$ *p.p., then*

$$\int_E f = \lim \int_E f_n.$$

A variant of the theorem of dominated convergence which is repeatedly appealed to in practice is

(E) *Let the series* $\Sigma u_n(x)$ *be boundedly convergent to sum* $s(x)$ *(i.e. if* $s_n(x) = \sum_1^n u_r(x)$, *then* $|s_n(x)| \leqslant M$). *Let* $\phi(x)$ *be integrable. Then*

$$\Sigma \int u_n \phi = \int s\phi.$$

D

Another test for the inversion of summation and integration which is convenient of application is the following extension of (B).

(F) *If either* $\int \Sigma |u_n|$ *or* $\Sigma \int |u_n|$ *is finite, then*

$$\int \Sigma u_n = \Sigma \int u_n.$$

That the alternative hypotheses are equivalent is shown by (B) itself.

Suppose that $\int \Sigma |u_n|$ is finite.

Then $\Sigma |u_n(x)|$ converges for almost all x.

So $\Sigma u_n(x)$ converges for almost all x, and it is dominated by the function $\Sigma |u_n(x)|$. The theorem of dominated convergence gives the result.

EXAMPLES ON CHAPTER III

(1) Show that the theorem of bounded convergence applies to $f_n(x) = nx/(1+n^2x^2)$, and the theorem of dominated convergence to $f_n(x) = n^{3/2}x/(1+n^2x^2)$, for $0 \leqslant x \leqslant 1$. [As these sequences do not converge uniformly, the standard criterion for $(R)\int f_n \to (R)\int f$ does not apply.]

(2) Give an example in which the sign $<$ is to be taken in Fatou's lemma, § 3·9 (v).

(3) Use Egoroff's theorem (Ch. II, Ex. 4) to prove the theorem of bounded convergence.

(4) Prove that, if $\alpha > 0$,

$$\lim_{n \to \infty} \int_0^n \left(1 - \frac{x}{n}\right)^n x^{\alpha-1}\,dx = \int_0^\infty e^{-x}x^{\alpha-1}dx.$$

(This is the main step in the proof that the integral and product definitions of the Γ function are equivalent. See, e.g., Whittaker and Watson, *Modern Analysis*, p. 235.)

(5) If $u_n(x) = ae^{-nax} - be^{-nbx}$ $(0 < a < b)$, prove that

$$\Sigma \int_0^\infty u_n(x)\,dx \neq \int_0^\infty \{\Sigma u_n(x)\}dx.$$

Verify directly that $\Sigma \int_0^\infty |u_n(x)|\,dx$ diverges.

By the appropriate expansions and term-by-term integrations, establish the following results, (6)–(8), stating to which of the theorems (A)–(F) of § 3·10 you appeal.

(6) $\displaystyle\int_0^1 \frac{x^p}{1-x} \log\left(\frac{1}{x}\right) dx = \frac{1}{(p+1)^2} + \frac{1}{(p+2)^2} + \cdots$ $(p > -1)$.

(7) $\displaystyle\int_0^\infty \frac{\sinh bx}{\sinh ax}\,dx = 2b\left\{\frac{1}{a^2 - b^2} + \frac{1}{(3a)^2 - b^2} + \frac{1}{(5a)^2 - b^2} + \cdots\right\}$

$$= \frac{\pi}{2a} \tan\frac{\pi b}{2a} \quad (0 < b < a).$$

(8) $\displaystyle\int_0^\infty \frac{\sin ax}{e^x - 1}dx = \pi\left(\frac{1}{e^{2\pi a} - 1} - \frac{1}{2\pi a} + \frac{1}{2}\right)$ $(a$ real$)$.

(9) For a given $f(x)$, define $M(x)$ as

$$\lim_{\delta \to 0} \{\text{upper bound of } f(x) \text{ in } (x - \delta, x + \delta)\},$$

and $m(x)$ as the limit of the lower bound.
With the notation of § 3·2, prove that, if $f(x)$ is bounded,

$$J = (L)\int_a^b M(x)\,dx, \quad j = (L)\int_a^b m(x)\,dx.$$

(10) Deduce from Example 9 that a necessary and sufficient condition that $f(x)$ should be R-integrable is that it is continuous p.p.

(11) Let $f(x)$ be L-integrable in (a, b). Prove that, given ϵ, there is a continuous function $\phi(x)$ such that

$$\int_a^b |f(x) - \phi(x)|\,dx < \epsilon.$$

(12) Prove that $\displaystyle\int_a^b |f(x+h) - f(x)|\,dx$ tends to 0 as h tends to 0.

CHAPTER IV

DIFFERENTIATION AND INTEGRATION

4·1. Differentiation and integration as inverse processes. If $f(x)$ is continuous and

$$F(x) = \int_a^x f(t)\, dt,$$

then it is well known that $F'(x) = f(x)$.

Moreover, if it is given that $\Phi(x)$ has a continuous derivative $\phi(x)$, then

$$\Phi(b) - \Phi(a) = \int_a^b \phi(x)\, dx.$$

We shall investigate how far these results hold for Lebesgue integrals.

If no assumption is made about $f(x)$ beyond its integrability (L), and $F(x)$ is its indefinite integral, then $F'(x)$ may fail to be equal to $f(x)$ in a set of measure zero. (For if E is any set of measure zero, and $f(x)$ is its characteristic function (see § 3·2), then $F'(x) = 0$ everywhere and $F'(x) \neq f(x)$ in E.) We shall prove in § 4·7 that $F'(x) = f(x)$ p.p.

4·2. The derivates of a function. As there is not in general a unique limit, as h tends to zero, of the ratio

$$\{\phi(x+h) - \phi(x)\}/h,$$

we consider

$$\overline{\lim_{h \to +0}}, \quad \varliminf_{h \to +0}, \quad \overline{\lim_{h \to -0}}, \quad \varliminf_{h \to -0}$$

of this ratio, these upper and lower limits always existing as finite numbers or $+\infty$ or $-\infty$. They are called the upper and lower *derivates* (or derived numbers) on the right and on the left respectively, and written as

$$D^+\phi(x), \quad D_+\phi(x), \quad D^-\phi(x), \quad D_-\phi(x),$$

the sign of the suffix showing whether h tends to 0 through positive or negative values and its position indicating an upper or a lower limit.

It is necessary and sufficient for the existence of the derivative $\phi'(x)$ that the four derivates should be equal.

It can be proved that the derivates of any measurable function are measurable. We confine ourselves to two special cases, which are sufficient for our purposes and which admit of simpler proofs.

If $\phi(x)$ is either (i) monotonic or (ii) continuous, then its derivates are measurable.

It is sufficient to give the detail of the proof for $D^+\phi(x)$.

Write $M(x, \delta)$ and $M_r(x, \delta)$ for the upper bounds of

$$\{\phi(x+h) - \phi(x)\}/h$$

for $0 < h < \delta$, when h takes respectively all values and only rational values.

$D^+\phi(x)$ is then $\lim_{\delta \to 0} M(x, \delta)$.

We shall prove that $M(x, \delta) = M_r(x, \delta)$. It is clear that $M_r \leqslant M$. To prove the reverse inequality, suppose that μ is any number less than $M(x, \delta)$. Then, for some positive h less than δ,

$$\frac{\phi(x+h) - \phi(x)}{h} > \mu.$$

If $\phi(x)$ is an increasing function, take a sequence of rationals r approaching h from above. Then

$$\frac{\phi(x+r) - \phi(x)}{r} \geqslant \frac{\phi(x+h) - \phi(x)}{r}$$

and the latter quotient is greater than μ if r is near enough to h.

If $\phi(x)$ is a decreasing function, the same inequality is obtained by letting r tend to h from below. In either case $M_r(x, \delta) \geqslant \mu$.

If $\phi(x)$ is assumed continuous instead of monotonic, $\phi(x+r) \to \phi(x+h)$ as $r \to h$, and so, again, $M_r(x, \delta) \geqslant \mu$.

Since μ is any number less than $M(x, \delta)$ it follows that $M_r(x, \delta) \geqslant M(x, \delta)$. Hence $M_r = M$.

For each fixed r, $\{\phi(x+r) - \phi(x)\}/r$ is measurable, and so also— from § 2·12 (v) and (vi)—are $M_r(x, \delta)$ and $\lim_{\delta \to 0} M_r(x, \delta)$.

Examples

(1) Evaluate the derivates at $x = 0$ of $\phi(x) = x \sin (1/x)$ $(x \neq 0)$, $\phi(0) = 0$.

(2) Construct a function having given numbers as derivates for a given value of x.

4·3. Vitali's covering theorem. Let E be a set of points x and Σ a set of intervals I. Suppose that, given any x of E and any ϵ, we can find an interval I of Σ, with length less than ϵ, having x as an interior or end-point. We shall then say that Σ *covers* E *in the sense of Vitali.* It is clear that, if O is any open set containing E, the subset of intervals of Σ contained in O also covers E in the sense of Vitali.

VITALI'S THEOREM. *Let E be a set of points of finite outer measure, and Σ a set of closed intervals I covering E in the sense of Vitali. Then we can find an enumerable set of intervals $I_1, I_2, \ldots$ of Σ, non-overlapping, such that the set of points of E not in any I_n has measure zero.*

We shall define the intervals $I_1, I_2, \ldots$ inductively.

Let O be an open set of finite measure containing E. Then we need retain only intervals of Σ contained in O.

Choose I_1 to be any interval of Σ. Suppose that $I_1, \ldots, I_n$ have been defined. Let k_n be the upper bound of the lengths of intervals of Σ which have no point in common with $\sum_1^n I_i$. Choose for I_{n+1} any such interval, of length greater than $\frac{1}{2}k_n$. This choice can always be made unless, for some n,

$$I_1 + \ldots + I_n \supset E,$$

in which case the theorem is proved.

It will be shown that the set R of points of E not in $\sum_1^\infty I_n$ has measure zero. Suppose on the contrary that $m^*R > 0$.

Let l_n be the length of I_n. Denote by J_n the interval of length $5l_n$ concentric with I_n.

Since $\sum_1^\infty I_n \subset O$, $\sum_1^\infty l_n$ converges and we can choose N such that

$$\sum_{N+1}^\infty mJ_n = 5\sum_{N+1}^\infty l_n < m^*R,$$

and so there are points of R which belong to no J_n for $n > N$. Let x be such a point.

Since x belongs to no I_n, it belongs to an I, of length l say, of Σ such that $II_n = 0$ for $n = 1, ..., N$.

But I has points in common with an I_n for some $n > N$, for if not, $l \leqslant k_n < 2l_{n+1}$ for every n. Since $\lim l_n = 0$, this is impossible. Let n_0 be the smallest value of n for which I and I_n have common points.

Then $\qquad\qquad\qquad l \leqslant k_{n_0-1}.$

But $n_0 > N$ and so, from the definition of x, x does not belong to J_{n_0}.

Since I contains both a point of I_{n_0} and a point not belonging to J_{n_0},

$$l \geqslant 2l_{n_0} > k_{n_0-1},$$

and this contradicts the preceding inequality.

Hence $mR = 0$ and Vitali's theorem is proved.

We add two corollaries, the first of which embodies the form of the theorem most useful in applications.

COROLLARY 1. *Under the hypotheses of Vitali's theorem, given ϵ, we can find a finite number of disjoint intervals $I_1, ..., I_n$ of Σ such that the outer measure of the set of points of E not covered by them is less than ϵ.*

COROLLARY 2. *The theorem (and the proof) hold in two (or more) dimensions if intervals are interpreted as squares (cubes, etc.).*

4·4. Differentiability of a monotonic function. The object of the next two theorems is to prove that a monotonic function has almost everywhere a finite derivative. We shall assume the function $\phi(x)$ to be increasing.

The set of values of x for which one of the upper derivates of $\phi(x)$ is $+\infty$ has measure zero.

Let E be the set in (a, b) at which $D^+\phi = +\infty$ or $D^-\phi = +\infty$, and suppose $mE = k$. Let K be any (large) number.

With each point x of E can be associated a sequence of intervals for which $\Delta\phi > K\Delta x$ (where $\Delta\phi$ is written for $\phi(x+h) - \phi(x)$).

By Vitali's covering theorem a finite number of these intervals can be selected, non-overlapping, and of measure greater than $\frac{1}{2}k$.

Summing over these intervals we have

$$\Sigma\Delta\phi > \tfrac{1}{2}Kk,$$

or, since ϕ is increasing,

$$\phi(b) - \phi(a) > \tfrac{1}{2}Kk.$$

This is false for sufficiently large K unless $k = 0$.

The set of points at which an upper derivate of an increasing function is greater than a lower derivate has measure zero.

Consider for definiteness $D^+\phi$ and $D_-\phi$ and suppose that the set of values of x for which $D^+\phi > D_-\phi$ has measure greater than zero.

This set is the sum of sets $E(u, v)$ in which

$$D^+\phi > u > v > D_-\phi$$

and u, v are rational numbers. There is then a pair (u, v) for which $E(u, v)$, $= E$ say, has measure k, greater than 0.

Enclose E in an open set O of measure less than $k + \epsilon$.

Any point x of E is the right-hand end-point of arbitrarily small intervals $(x-h, x)$ for which $\Delta\phi = \phi(x) - \phi(x-h) < vh$.

Vitali's theorem enables us to pick out a finite number of these intervals $I_1, \ldots, I_m$, disjoint and containing a part E_1 of E of measure at least $k - \epsilon$ and contained in O. Summing over these intervals, we have $\Sigma\Delta\phi < v(k + \epsilon)$.

Each point of E_1 is the left-hand end-point of intervals for which $\phi(x+h) - \phi(x) > uh$. By Vitali's theorem we can pick out from them disjoint intervals $J_1, ..., J_n$, contained in $I_1 + ... + I_m$ and of total measure greater than $k - 2\epsilon$.

Summing over these intervals, we have $\Sigma\Delta\phi > u(k - 2\epsilon)$.

But, since ϕ is an increasing function, the sum of its increments over the I's is not less than over the J's. Hence

$$v(k + \epsilon) > u(k - 2\epsilon).$$

If ϵ is small enough, this contradicts the hypothesis $u > v$.

This argument is applicable to any pair of upper and lower derivates on the same or opposite sides.

The two theorems of this section together show that, if $\phi(x)$ is an increasing function, a finite $\phi'(x)$ exists almost everywhere. The same result holds for a decreasing function.

4·5. The integral of the derivative of an increasing function.

If $\phi(x)$ is increasing in (a, b), then $\phi'(x)$ is integrable and

$$\int_a^b \phi'(x)\,dx \leqslant \phi(b) - \phi(a).$$

$\phi'(x)$ may be undefined (or infinite) in a set of measure zero, but, from § 3·8 (i), this will not affect the integral.

Let h take a decreasing sequence of values tending to 0.

For values of x greater than b, it will be convenient to suppose, as we may, that $\phi(x)$ is equal to $\phi(b)$.

Write

$$\psi_h(x) = \frac{\phi(x+h) - \phi(x)}{h}.$$

From § 4·4, as h tends to zero, $\psi_h(x)$ tends to $\phi'(x)$ p.p.

Since $\psi_h(x) \geqslant 0$ for all x and h, we may apply Fatou's lemma (§ 3·9) and obtain

$$\int_a^b \phi'(x)\,dx \leqslant \varliminf \int_a^b \psi_h(x)\,dx.$$

But $\displaystyle\int_a^b \psi_h(x)\,dx = \frac{1}{h}\left\{\int_{a+h}^{b+h}\phi(x)\,dx - \int_a^b\phi(x)\,dx\right\}$

$\displaystyle = \frac{1}{h}\left\{\int_b^{b+h}\phi(x)\,dx - \int_a^{a+h}\phi(x)\,dx\right\}.$

As h tends to 0, $\displaystyle\lim\frac{1}{h}\int_b^{b+h}\phi(x)\,dx = \phi(b)$ and, since

$\displaystyle\int_a^{a+h}\phi(x)\,dx \geqslant h\phi(a)$, we have $\displaystyle\underline{\lim}\,\frac{1}{h}\int_a^{a+h}\phi(x)\,dx \geqslant \phi(a)$.

This proves the theorem.

It is interesting to construct a continuous function $\phi(x)$ for which the sign $<$ is to be taken in the theorem.

Let E be Cantor's ternary set in $(0,1)$ (Ch. I, Ex. 7). In the intervals of CE (and at their end-points) define $\phi(x)$ as follows. For $\frac{1}{3}\leqslant x \leqslant \frac{2}{3}$, $\phi(x) = \frac{1}{2}$. For $\frac{1}{9}\leqslant x \leqslant \frac{2}{9}$, $\phi(x) = \frac{1}{4}$, and for $\frac{7}{9}\leqslant x \leqslant \frac{8}{9}$, $\phi(x) = \frac{3}{4}$. For $\frac{1}{27}\leqslant x \leqslant \frac{2}{27}$, $\phi(x) = \frac{1}{8}$; for $\frac{7}{27}\leqslant x \leqslant \frac{8}{27}$, $\phi(x) = \frac{3}{8}$; and so on, repeatedly trisecting along Ox and bisecting along Oy. (A diagram will help the reader.) At a point x of E which is not an end-point of an interval of CE, $\phi(x)$ can be uniquely defined as the limit of values of ϕ taken in a sequence of intervals approaching x, and $\phi(x)$ is continuous at all points of $(0,1)$.

Since $\phi'(x) = 0$ at all points within an interval of CE and the sum of the lengths of these intervals is 1, $\phi'(x) = 0$ p.p., and so $\displaystyle\int_0^1 \phi'(x)\,dx = 0$. But $\phi(1) - \phi(0) = 1$.

4·6. Functions of bounded variation. Let $f(x)$ be defined for $a \leqslant x \leqslant b$. Take a set of points of division

$$a = x_0 < x_1 < \ldots < x_k = b.$$

Let p be the sum of those differences $f(x_{r+1}) - f(x_r)$ for $r = 0, 1, \ldots, k-1$, which are positive, and $-n$ the sum of those which are negative. Then

$$f(b) - f(a) = p - n$$

and $\displaystyle\sum_{r=0}^{k-1}|f(x_{r+1}) - f(x_r)| = p + n = t,$ say.

Suppose that P, N, T are the upper bounds of p, n, t for all modes of subdivision of (a, b). It is plain that

$$P(\text{or } N) \leqslant T \leqslant P + N.$$

Call the numbers P, N, T respectively the positive, negative and total variations of $f(x)$ in (a, b). Either these three numbers are all finite or T and at least one of P and N are infinite; if T is finite we say that $f(x)$ *has bounded variation* in (a, b), and that T is its *total variation*.

If $f(x)$ has bounded variation, $T = P + N$. Whatever the mode of subdivision, we have

$$p = n + f(b) - f(a),$$
$$\leqslant N + f(b) - f(a).$$

This being true for all values of p, we deduce that

$$P \leqslant N + f(b) - f(a),$$
or $$P - N \leqslant f(b) - f(a).$$

A similar argument gives

$$N - P \leqslant f(a) - f(b),$$
and so $$P - N = f(b) - f(a).$$

Then $\quad T \geqslant p + n = p + p - \{f(b) - f(a)\} = 2p + N - P,$

and since this is true for any choice of points of division, we may replace p by its upper bound P, giving

$$T \geqslant P + N.$$

This combined with the obvious inequality

$$T \leqslant P + N$$

gives the result.

A function of bounded variation is the difference between two bounded increasing functions.

Let $f(x)$ have bounded variation in (a, b) and $a \leqslant x \leqslant b$. Let $P(x)$ and $N(x)$ be the positive and negative variations of $f(x)$ in the interval (a, x). The proof of the last theorem shows that

$$f(x) = \{f(a) + P(x)\} - N(x)$$

and the right-hand side is the difference between two bounded increasing functions of x.

Conversely, if $f(x) = \phi(x) - \psi(x)$, where $\phi(x)$ and $\psi(x)$ are bounded increasing functions, it follows from summing the inequalities

$$|f(x_{r+1}) - f(x_r)| \leqslant \{\phi(x_{r+1}) - \phi(x_r)\} + \{\psi(x_{r+1}) - \psi(x_r)\}$$

that $f(x)$ has bounded variation.

Examples

(1) Prove that the sum and product of two functions of bounded variation have bounded variation.

(2) Prove that, if $f(x)$ is of bounded variation and continuous, the functions $P(x)$, $N(x)$ and $T(x)$ are continuous.

(3) Prove that the functions $x \sin(1/x)$, $x^2 \sin(1/x^2)$ (defined to be 0 for $x = 0$) are not of bounded variation in any interval containing $x = 0$. Prove that $x^2 \sin(1/x^{3/2})$ has bounded variation.

(4) Prove that a necessary and sufficient condition for the curve $x = x(t)$, $y = y(t)$, $a \leqslant t \leqslant b$, to have finite length is that $x(t)$ and $y(t)$ have bounded variation.

4·7. Differentiation of the indefinite integral.

If $F(x) = F(a) + \int_a^x f(t)\,dt$, *then* $F'(x) = f(x)$ *almost everywhere.*

$F(x)$, being the integral of $f_+ - f_-$, where f_+ and f_- are the positive functions defined in § 3·1, is the difference between two increasing functions. Therefore $F'(x)$ exists p.p., and it remains to show that $F'(x) = f(x)$ p.p.

Suppose first that $f(x)$ is bounded, say $|f(x)| \leqslant K$.

Let h take a sequence of values tending to 0.

Then
$$\left| \frac{F(x+h) - F(x)}{h} \right| = \left| \frac{1}{h} \int_x^{x+h} f(t)\,dt \right| \leqslant K$$

and
$$\frac{F(x+h) - F(x)}{h} \to F'(x) \quad \text{p.p.}$$

By the theorem of bounded convergence, if c is any point of (a, b),

$$\int_a^c F'(x)\,dx = \lim_{h\to 0}\frac{1}{h}\int_a^c \{F(x+h)-F(x)\}\,dx$$

$$= \lim_{h\to 0}\left\{\frac{1}{h}\int_c^{c+h} F(x)\,dx - \frac{1}{h}\int_a^{a+h} F(x)\,dx\right\}$$

$$= F(c)-F(a),$$

on account of the continuity of F.

Hence
$$\int_a^c \{F'(x)-f(x)\}\,dx = 0$$

for all values of c in (a, b).

From § 3·8 (iii), $F'(x)=f(x)$ p.p.

Now suppose that $f(x)$ is unbounded. By the usual decomposition of f into its positive and negative parts $(f = f_+ - f_-)$, it is sufficient to give the proof for positive $f(x)$.

Let $\{f(x)\}_n$ be defined as in § 3·5.

Then

$$\int_a^x [f(t)-\{f(t)\}_n]\,dt,$$

being the integral of a positive function, increases with x. Therefore its derivative exists p.p. and is never negative.

By the result for bounded functions

$$\frac{d}{dx}\int_a^x \{f(t)\}_n\,dt = \{f(x)\}_n \quad \text{p.p.}$$

Hence
$$F'(x)\geqslant \{f(x)\}_n \quad \text{p.p.}$$

or, since n is arbitrarily large,

$$F'(x)\geqslant f(x) \quad \text{p.p.}$$

From this,

$$\int_a^b F'(x)\,dx \geqslant \int_a^b f(x)\,dx = F(b)-F(a).$$

But the theorem of § 4·5 gives the reverse of this inequality, and so

$$\int_a^b \{F'(x) - f(x)\} dx = 0.$$

Since the integrand is p.p. greater than or equal to zero, it is, by § 3·8 (ii), zero p.p., and the theorem is proved.

As the example given in § 4·1 shows, there is a sense in which this theorem is the best possible.

The theorem contains as a special case a fundamental metric property of sets of points. If E is measurable we define the *average density* of E in an interval I to be $m(EI)/mI$.

The upper and lower right-hand densities of E at a point x are the upper and lower limits of the average density in $(x, x+h)$ as $h \to +0$. Similarly the left-hand densities are the limits of the average density in $(x-h, x)$. If all these four numbers are equal we speak simply of the *density* of E at x.

By applying the theorem of this section to the integral of the characteristic function of E, we have:

The density of a measurable set E is 1 at almost all points of E, and is 0 at almost all points of CE.

4·8. Absolutely continuous functions. Let $f(x)$ be defined for $a \leqslant x \leqslant b$. Suppose that $(x_1, x_1 + h_1), \ldots, (x_n, x_n + h_n)$ are non-overlapping intervals in (a, b). If, given ϵ, we can find δ such that

$$\sum_{r=1}^n |f(x_r + h_r) - f(x_r)| < \epsilon$$

for all choices of intervals with $\Sigma h_r < \delta$, $f(x)$ is said to be *absolutely continuous*.

By taking *one* interval $(x, x+h)$, we see that an absolutely continuous function is continuous.

It is clear that the sum of two absolutely continuous functions is absolutely continuous.

An absolutely continuous function has bounded variation. Given any subdivision of (a, b), we can, by inserting fresh points of division, split up (a, b) into N sets of intervals, each set being of total length less than δ, where $N \leqslant \dfrac{b-a}{\delta} + 1$. Then $\Sigma \, | \, \Delta f \, |$, summed over all these sub-intervals, and *a fortiori* over the original sub-intervals, is less than $N\epsilon$.

The importance of the class of absolutely continuous functions is that it is the same as the class of indefinite Lebesgue integrals. We proceed to prove this.

LEMMA. *If $\phi(x)$ is absolutely continuous in an interval, and $\phi'(x) = 0$ p.p., then $\phi(x)$ is constant.*

The lemma will follow if, for any interval (a, b) in which the hypotheses hold, $\phi(b) = \phi(a)$.

Let E be the set, of measure $b - a$, in which $\phi'(x) = 0$.

With each point x of E is associated an interval $(x, x + h)$ (arbitrarily small) such that

$$| \, \phi(x + h) - \phi(x) \, | < \eta h,$$

where η is any positive number.

Given ϵ, we can by Vitali's theorem (§ 4·3) pick out a finite set, J_1 say, of these intervals, non-overlapping, such that the measure of the complementary intervals (J_2) is less than ϵ. Then

$$| \, \phi(b) - \phi(a) \, | \leqslant (\Sigma_1 + \Sigma_2) \, | \, \phi(x + h) - \phi(x) \, |,$$

where Σ_1 and Σ_2 denote summations over the intervals of J_1 and J_2 respectively. Now $0 \leqslant \Sigma_1 \leqslant \eta(b - a)$. And, by absolute continuity, Σ_2 tends to 0 as ϵ tends to 0. Hence

$$| \, \phi(b) - \phi(a) \, | \leqslant \eta(b - a).$$

Since η is arbitrarily small, $\phi(b) = \phi(a)$.

THEOREM. *A necessary and sufficient condition that a function should be an indefinite integral is that it should be absolutely continuous.*

We shall show that, for any set E of sufficiently small measure contained in E_0, $\displaystyle\int_E f(x)\,dx$ can be made arbitrarily small. The special case in which E is the sum of intervals gives the necessity of the condition in the theorem.

If $f(x)$ is bounded, the first mean-value theorem (§ 3·7 (i)) gives the result. For unbounded f, since $\left|\int f\right| \leqslant \int |f|$, we may suppose f positive. Then, given ϵ, we can find n such that

$$\int_{E_\bullet} f \leqslant \int_{E_\bullet} (f)_n + \tfrac{1}{2}\epsilon.$$

Take $\delta = \epsilon/2n$. Then, for any E with measure less than δ,

$$\int_E f < nmE + \tfrac{1}{2}\epsilon < \epsilon.$$

We have now to prove the sufficiency half of the theorem, i.e. that, if $\psi(x)$ is absolutely continuous, $\psi'(x)$ is integrable

and
$$\psi(x) - \psi(a) = \int_a^x \psi'(x)\,dx.$$

$\psi(x)$, having bounded variation, is the difference between two increasing functions

$$\psi(x) = \psi_1(x) - \psi_2(x).$$

From § 4·5, $\psi_1'(x)$ and $\psi_2'(x)$, existing p.p. are integrable and so $\psi'(x)$ is integrable.

Then $\phi(x) = \psi(x) - \int_a^x \psi'(x)\,dx$ is the difference between two absolutely continuous functions and so is absolutely continuous.

Also, from § 4·7, $\phi'(x) = 0$ p.p.

Hence, by the lemma of this section, $\phi(x)$ is constant and so

$$\psi(x) - \psi(a) \stackrel{\backprime}{=} \int_a^x \psi'(x)\,dx,$$

and the theorem is proved.

We now see that the continuous increasing function $\phi(x)$ defined at the end of § 4·5 is not absolutely continuous in (0, 1) since

$$\phi(1) - \phi(0) > \int_0^1 \phi'(x)\,dx.$$

EXAMPLES ON CHAPTER IV

(1) Prove that the product of two absolutely continuous functions is absolutely continuous.

(2) By calculating the sum of its increments over suitable intervals, give a direct proof that the function defined in § 4·5 is not absolutely continuous.

(3) Construct a direct proof of the density theorem (end of § 4·7) without using properties of integrals.

(4) From the density theorem deduce the $F' = f$ theorem for bounded f.

(5) Prove that the Lebesgue integral will integrate any bounded derivative. (Ex. 3 of § 3·5 shows that the hypothesis of boundedness cannot be omitted.)

(6) Prove that any increasing function $f(x)$ can be decomposed into the sum of three functions

$$\phi(x) + \psi(x) + \chi(x),$$

where $\phi(x)$ is absolutely continuous, $\psi(x)$ is continuous with $\psi'(x) = 0$ p.p., $\chi(x) = \sum_{\xi_r < x} \{f(\xi_r + 0) - f(\xi_r - 0)\}$ summed over discontinuities ξ_r.

E

FURTHER PROPERTIES OF THE INTEGRAL

This chapter contains theorems of a familiar type used in working with integrals. It will be noted that the conditions under which they can be established for Lebesgue integrals are very wide.

5·1. Integration by parts. *If $F(x)$, $G(x)$ are respectively indefinite integrals of $f(x), g(x)$, then*

$$\int_a^b Fg\,dx = [FG]_a^b - \int_a^b fG\,dx.$$

From Example 1 of Chapter IV the product of the absolutely continuous functions F, G is absolutely continuous. From § 4·8, FG is the integral of its derivative, which is equal to $Fg + fG$ almost everywhere. (Since F, G are continuous, it is easy to see that Fg and fG are integrable. See example of § 3·7.)

5·2. Change of variable. *Let $x = x(t)$ be a non-decreasing absolutely continuous function of t, with $x(\alpha) = a$, $x(\beta) = b$. Then, if $f(x)$ is integrable,*

$$\int_a^b f(x)\,dx = \int_\alpha^\beta f\{x(t)\}x'(t)\,dt.$$

Let $F(x) = \int_a^x f(x)\,dx$. We start by proving two lemmas.

LEMMA 1. *If $F(x)$, $x(t)$ are absolutely continuous functions of x and t respectively and $x(t)$ is monotonic, then $F\{x(t)\}$ is an absolutely continuous function of t.*

For, if t_r, t'_r is a typical interval of a set contained in (α, β) and $x_r = x(t_r)$, $x'_r = x(t'_r)$, then the absolute continuity of $x(t)$ implies that, as $\Sigma(t'_r - t_r)$ tends to zero, so does $\Sigma(x'_r - x_r)$. But $\Sigma \,|\, F(x'_r) - F(x_r)\,|$ tends to zero as $\Sigma(x'_r - x_r)$ tends to zero, and this proves Lemma 1.

LEMMA 2. *Let $x(t)$ be a non-decreasing absolutely continuous function. Let X be a set of values of x, and T the set of values of t for which $x = x(t)$ is in X. Then $mX = 0$ implies that $x'(t) = 0$ p.p. in T.*

With the notation of Lemma 1, since $x(t)$ is absolutely continuous,

$$x'_r - x_r = \int_{t_r}^{t'_r} x'(t)\,dt.$$

Hence, if O_x, O_t are corresponding open sets of which typical intervals are respectively (x_r, x'_r) and (t_r, t'_r),

$$mO_x = \int_{O_t} x'(t)\,dt.$$

Let $O_{x,n}$ be a decreasing sequence of open sets containing X such that $\lim_{n \to \infty} mO_{x,n} = 0$.

Then the product of the corresponding sets $O_{t,n}$, say T_0, contains T and

$$\int_{T_0} x'(t)\,dt \leqslant \int_{O_{t,n}} x'(t)\,dt = mO_{x,n} < \epsilon \quad \text{for } n > n_0(\epsilon).$$

Therefore $$\int_{T_0} x'(t)\,dt = 0.$$

Since $x'(t) \geqslant 0$, we have, from § 3·8 (ii), $x'(t) = 0$ p.p. in T_0 and so p.p. in T, and the lemma is proved.

We now prove the main theorem.

At almost all points of (α, β) a finite derivative $x'(t)$ exists. Let T_1 be the set of t for which $x'(t) \neq 0$.

We have, if $x(t+h) \neq x(t)$,

$$\frac{F\{x(t+h)\} - F\{x(t)\}}{h} = \frac{F\{x(t+h)\} - F\{x(t)\}}{x(t+h) - x(t)} \, \frac{x(t+h) - x(t)}{h}.$$

As h tends to 0, the second quotient on the right-hand side tends to $x'(t)$ for almost all values of t.

The first quotient on the right-hand side has the limit $f\{x(t)\}$ for all x except a set of measure zero, say X.

By Lemma 2 the subset of T_1 for which $x(t)$ is in X has measure zero.

Hence, p.p. in T_1,

$$\frac{d}{dt}F\{x(t)\} = f\{x(t)\}x'(t).$$

Suppose first that $f(x)$ is bounded, say $|f(x)| \leqslant K$. Then

$$|F\{x(t+h)\} - F\{x(t)\}| \leqslant K\{x(t+h) - x(t)\}.$$

So $x'(t) = 0$ implies that $\dfrac{d}{dt}F\{x(t)\} = 0$.

We have thus shown that, if $f(x)$ is bounded,

$$\frac{d}{dt}F\{x(t)\} = f\{x(t)\}x'(t) \quad \text{p.p. in } (\alpha, \beta).$$

It follows from Lemma 1 and the theorem of § 4·8 that

$$F(b) - F(a) = \int_\alpha^\beta f\{x(t)\}x'(t)\,dt.$$

To establish the result for an unbounded $f(x)$, it is sufficient to give the proof for $f(x)$ positive.

Let
$$f_n(x) = f(x) \quad \text{if } f(x) \leqslant n,$$
$$f_n(x) = n \quad\;\; \text{if } f(x) > n,$$

and let
$$F_n(x) = \int_a^x f_n(x)\,dx.$$

Then
$$F(b) - F(a) = \lim_{n \to \infty}\{F_n(b) - F_n(a)\},$$

by the definition of the integral of an unbounded function, and the right-hand side (by what we have just proved for bounded functions) is equal to

$$\lim_{n \to \infty}\int_\alpha^\beta f_n\{x(t)\}x'(t)dt.$$

As n increases and tends to infinity $f_n\{x(t)\}\,x'(t)$ increases and tends to $f\{x(t)\}x'(t)$.

But $f\{x(t)\}x'(t)$ is integrable in (α,β)—for it is integrable over the set T_1, being p.p. the derivative of the absolutely continuous function $F\{x(t)\}$, and it vanishes p.p. over the set complementary to T_1.

By Beppo Levi's theorem § 3·10 (A),

$$\lim_{n\to\infty}\int_\alpha^\beta f_n\{x(t)\}\,x'(t)\,dt = \int_\alpha^\beta f\{x(t)\}\,x'(t)\,dt,$$

and the left-hand side has been proved equal to $F(b)-F(a)$.

5·3. Multiple integrals.

The integral $\displaystyle\int_E f(x)\,dx$ was defined as the plane-measure of an ordinate-set in two dimensions. In the same way the value of $\displaystyle\iint_E f(x,\,y)dxdy$ is defined to be the three-dimensional measure of the appropriate ordinate-set erected on the plane set E; and so on for higher dimensions.

The mode of evaluation of a double integral at the elementary level (e.g. when $f(x,y)$ is continuous) is by repeated integration with respect to each variable in turn. We shall give a general theorem of this kind for a Lebesgue multiple integral, which shows that the existence of the multiple integral carries with it without further condition the existence p.p. of the repeated integrals with respect to the separate variables. The method will be clear if we give the detail when the number of variables of integration is two.

LEMMA. *Let E be a measurable plane set. Let E_x be the linear set which is the section of E by the ordinate distant x from the origin. Then E_x is measurable for almost all x and*

$$\mu E = \int mE_x\,dx,$$

the limits of integration being such as to include the whole of E.

The lemma is plainly true if E is a rectangle, or a finite number of rectangles. We shall prove it for an open set O.

By the network construction (§ 1·5), $O = \lim G_n$, where $G_n \subset G_{n+1}$ and G_n consists of a finite number of rectangles. Also $O_x = \lim_{n \to \infty} (G_n)_x$.

Then $\mu O = \lim_{n \to \infty} \mu G_n$ (from the definition of μO)

$$= \lim_{n \to \infty} \int m(G_n)_x dx = \int m O_x dx, \text{ by Beppo Levi's theorem, § 3·10}(A).$$

Thus the lemma holds for open sets and similarly (or by complements) for closed sets.

If now E is any measurable set, take a decreasing sequence $O_1, O_2, \ldots$ of open sets containing E such that $\mu O_n \to \mu E$; and an increasing sequence $Q_1, Q_2, \ldots$ of closed sets contained in E such that $\mu Q_n \to \mu E$.

Then, for each x, $\quad (O_n)_x \supset E_x \supset (Q_n)_x,$

$$\lim_{n \to \infty} \int m(O_n)_x dx = \lim_{n \to \infty} \mu O_n = \mu E$$

and

$$\lim_{n \to \infty} \int m(Q_n)_x dx = \lim_{n \to \infty} \mu Q_n = \mu E.$$

Therefore

$$\lim_{n \to \infty} \int \{m(O_n)_x - m(Q_n)_x\} dx = 0.$$

Since the integrand is a positive decreasing function of n, we have from § 3·9 (ii) and § 3·8 (ii),

$$\lim_{n \to \infty} \{m(O_n)_x - m(Q_n)_x\} = 0 \quad \text{p.p.}$$

Therefore E_x is measurable for almost all x and $\int m E_x dx$, lying between two sequences of integrals each tending to μE, is equal to μE. The lemma is proved.

We could prove similarly a three-dimensional form of the lemma, and this is what will be needed in dealing with double integrals. If E is a measurable set in three dimensions and E_x is the plane set which is its section by a plane perpendicular to the x-axis, distant x from the origin, then the measure of E is equal to $\int \mu E_x dx$.

We are now ready to state Fubini's theorem on multiple integrals and in doing so we refer to the discussion of § 3·6. The domain of integration may be any measurable bounded set in the plane, or it may be the whole plane provided that the integral is absolutely convergent. On this understanding the most convenient notation is to leave the ranges of integration not explicitly specified.

5·4. Fubini's theorem. *If* $\iint f(x,y)\,dx\,dy$ *exists, then* $\int f(x,y)\,dy$ *exists for almost all x, is an integrable function of x and*

$$\iint f(x,y)\,dx\,dy = \int dx \int f(x,y)\,dy.$$

Similarly,

$$\iint f(x,y)\,dx\,dy = \int dy \int f(x,y)\,dx.$$

For the proof, suppose first that f is positive. The three-dimensional form of the lemma just stated, applied to the ordinate-set of f, gives the result.

If f takes values of either sign, we have only to write $f = f_+ - f_-$, where f_+ and f_- are positive, and apply the lemma to the ordinate-sets of f_+ and f_-.

The operation involving multiple integrals which one most often wishes to justify in practice is the interchange of the order of integration in a repeated integral. The condition of validity of Fubini's theorem which necessitates a direct investigation of the multiple integral may not be easy to establish. The following simple variants are analogous to § 3·10 (B) and (F) for integration of series.

If $f(x, y)$ is a measurable function of (x, y) and if $f \geqslant 0$ for all (x, y) in the domain of integration, then the existence of any one of

$$\iint f \, dx \, dy, \qquad \int dx \int f \, dy, \qquad \int dy \int f \, dx$$

implies the existence and equality of all three.

Suppose that $\int dx \int f \, dy$ exists. Define $f_n(x, y)$ as follows:

$$f_n = f \text{ if } |x| \leqslant n, \ |y| \leqslant n \text{ and } f \leqslant n,$$
$$f_n = n \text{ if } |x| \leqslant n, \ |y| \leqslant n \text{ and } f > n,$$
$$f_n = 0 \text{ if } |x| > n \text{ or } |y| > n.$$

Then f_n is a bounded measurable function and $\iint f_n \, dx \, dy$ exists and is equal to either of the repeated integrals of f_n.

From the hypothesis, $\int f \, dy$ exists for almost all x. $\int f_n \, dy$ increases with n and tends to $\int f \, dy$.

By § 3·10 (A), $\qquad \int dx \int f_n \, dy \to \int dx \int f \, dy.$

But the left-hand side is $\iint f_n \, dx \, dy$ and its limit, as $n \to \infty$, is $\iint f \, dx \, dy$.

Hence $\iint f \, dx \, dy = \int dx \int f \, dy$, and the rest follows.

If $f(x, y)$ is a measurable function of (x, y) and if $\int dx \int |f| \, dy$ exists, then

$$\int dx \int f \, dy = \int dy \int f \, dx.$$

From the last result, $\iint |f| \, dx \, dy$ exists, therefore so does $\iint f \, dx \, dy$ and the conclusion follows from Fubini's theorem.

5·5. Differentiation of multiple integrals. This is an operation of theoretical interest rather than one which occurs in day-to-day analysis, and we only mention the main result (for double integrals).

If S is a square of side h, containing (x, y), then p.p.

$$\lim_{h \to 0} \frac{1}{h^2} \int\int_S f(x, y)\, dx\, dy = f(x, y).$$

A proof can be based on Vitali's covering theorem § 4·3 (note Corollary 2).

5·6. The class L^p. We say that $f(x)$ is in the Lebesgue class L^p (where $p > 0$) for a given set of values of x if $f(x)$ is measurable and $|f(x)|^p$ is integrable in the set. For example:

(1) $x^{-\frac{1}{2}}$ is in L^p, for $p < 2$, in $(0, 1)$.

(2) In a finite interval (a, b), a function in L^p is also in L^q for $0 \leqslant q < p$; a bounded function is in L^p for every p.

(3) The function $x^{-\frac{1}{2}}(1 + |\log x|)^{-1}$ is in L^2 in $(0, \infty)$, but not in L^p in $(0, \infty)$ for any value of p other than 2.

The most interesting case is $p \geqslant 1$ and we shall assume this. The integrals will be supposed to be taken over a given set E of finite or infinite measure.

If $p > 1$, the *conjugate* index p' is defined by

$$\frac{1}{p} + \frac{1}{p'} = 1 \quad \text{i.e.} \quad p' = \frac{p}{p-1}.$$

The classes $L^p, L^{p'}$ will be called conjugate; L^2 is self-conjugate. We define $N_p(f)$, the *norm* of f, by

$$N_p(f) = \left(\int_E |f|^p\, dx \right)^{1/p}.$$

HÖLDER'S INEQUALITY. *If, in E, f is in L^p and g is in L$^{p'}$, then*

$$\int |fg| \leqslant \left\{ \int |f|^p \right\}^{1/p} \left\{ \int |g|^{p'} \right\}^{1/p'},$$

with equality only when $A|f|^p = B|g|^{p'}$ p.p. for some constants A, B not both zero (or, as we may say, $|f|^p$ and $|g|^{p'}$ are effectively proportional).

Write

$$|f|^p = \phi, \quad |g|^{p'} = \psi, \quad \frac{1}{p} = \alpha, \quad \frac{1}{p'} = \beta.$$

Then we have to prove that

$$\int \phi^\alpha \psi^\beta \leqslant \left(\int \phi\right)^\alpha \left(\int \psi\right)^\beta, \quad (\alpha + \beta = 1).$$

By the inequality of the arithmetic and geometric means, i: $a > 0$, $b > 0$,

$$a^\alpha b^\beta \leqslant a\alpha + b\beta, \quad (\alpha + \beta = 1),$$

with equality only when $a = b$.

In the last inequality write

$$a = \frac{\phi}{\displaystyle\int_E \phi}, \quad b = \frac{\psi}{\displaystyle\int_E \psi}$$

and integrate over E.

The right-hand side is integrable and its value is $\alpha + \beta = 1$. The left-hand side, being measurable, is integrable and the result follows.

MINKOWSKI'S INEQUALITY. *If f and g are in L^p, then*

$$\left\{\int |f + g|^p\right\}^{1/p} \leqslant \left\{\int |f|^p\right\}^{1/p} + \left\{\int |g|^p\right\}^{1/p},$$

with equality only if f and g are effectively proportional.

For

$$\int |f + g|^p \leqslant \int |f| \cdot |f + g|^{p-1} + \int |g| \cdot |f + g|^{p-1}$$

$$\leqslant \left\{\int |f|^p\right\}^{1/p} \left\{\int |f + g|^p\right\}^{1/p'} + \left\{\int |g|^p\right\}^{1/p} \left\{\int |f + g|^p\right\}^{1/p'}$$

by Hölder's inequality.

Dividing each side by $\left\{\int |f + g|^p\right\}^{1/p'}$, we have the result.

5·7. The metric space L^p.

The members of a set are said to be the elements of a *metric space* if, for every pair x, y, there is defined a *distance-function* $d(x, y)$ with the properties:

(i) $\qquad\qquad d(x, y) > 0 \quad$ if $x \neq y; \quad d(x, x) = 0.$

(ii) $\qquad\qquad\qquad d(x, y) = d(y, x).$

(iii) $\quad d(x, z) \leqslant d(x, y) + d(y, z)$—the triangle inequality.

Functions of L^p are elements of a metric space if we take

$$d(f, g) = N_p(f - g).$$

Two functions differing only in a set of measure zero are indistinguishable as elements of the metric space. With this convention, the properties (i), (ii), (iii) of the distance function are satisfied, (iii) being Minkowski's inequality.

The reader will recognize in the following discussion an extension to the space of functions in L^p of ideas such as limit-point in the theory of sets of points.

If f_n and f are in L^p, and $N_p(f_n - f) \to 0$ as $n \to \infty$, we say that $f_n \to f(L^p)$ or alternatively that f_n *converges strongly* to f (with index p).

A necessary and sufficient condition that $f_n \to f(L^p)$ *is that* $N_p(f_m - f_n) \to 0$ *as m and n tend to infinity. Two such limit functions f can differ only in a set of measure zero. Moreover, there is a sub-sequence n_r such that* $f_{n_r} \to f$ *p.p.*

We first prove sufficiency. Given ϵ, there is $n_0(\epsilon)$ such that for $m \geqslant n_0$, $n \geqslant n_0$,

$$\int |f_m - f_n|^p \, dx < \epsilon^{p+1},$$

and so the set in which $|f_m - f_n| > \epsilon$ has measure less than ϵ.

Replace ϵ by $\epsilon/2, \dots, \epsilon/2^k, \dots$ successively and let $n_1, \dots, n_k, \dots$ be the indices corresponding to the n_0 of the last paragraph. Then

$$\sum_{k=r}^{\infty} |f_{n_{k+1}}(x) - f_{n_k}(x)| < \sum_{k=r}^{\infty} \frac{\epsilon}{2^k} = \frac{\epsilon}{2^{r-1}},$$

except in a set of measure at most $\epsilon/2^{r-1}$. Since the measure of the exceptional set tends to 0 as $r \to \infty$, it follows that the series

$$\sum_{k=0}^{\infty} \left\{ f_{n_{k+1}}(x) - f_{n_k}(x) \right\},$$

is absolutely convergent p.p., that is to say, there is a function $f(x)$ defined p.p. to which the sub-sequence $f_{n_r}(x)$ converges as $r \to \infty$.

We shall prove that f_n converges strongly to f with index p. By Fatou's lemma (§ 3·9),

$$\int |f - f_n|^p \, dx \leqslant \lim_{r \to \infty} \int |f_{n_r} - f_n|^p \, dx \leqslant \epsilon^{p+1} \quad \text{if } n \geqslant n_0,$$

i.e.
$$f_n \to f(L^p).$$

To prove the uniqueness of the limit function (ignoring differences in sets of measure zero), suppose that f_n also converged to $g(L^p)$. By Minkowski's inequality,

$$N_p(f - g) \leqslant N_p(f - f_n) + N_p(f_n - g),$$

and the right-hand side tends to zero as $n \to \infty$.

The necessity part of the theorem also follows at once from Minkowski's inequality. For

$$N_p(f_m - f_n) \leqslant N_p(f_m - f) + N_p(f - f_n).$$

EXAMPLES ON CHAPTER V

(1) Investigate the question of existence and equality of the double and repeated integrals of the following functions over the square $0 \leqslant x \leqslant 1$, $0 \leqslant y \leqslant 1$.

(i) $\quad \dfrac{x^2 - y^2}{(x^2 + y^2)^2},$ (ii) $\quad \dfrac{1}{(1 - xy)^\alpha},$

(iii) $\quad f(x, y) = \dfrac{1}{(x - \frac{1}{2})^3}$ for $0 < y < |x - \frac{1}{2}|$

$\qquad f(x, y) = 0$ for $y \geqslant |x - \frac{1}{2}|$ and $y = 0$.

(2) Prove that, if $f(x)$ is integrable, then

$$f_\alpha(x) = \frac{1}{\Gamma(\alpha)}\int_0^x (x-t)^{\alpha-1} f(t)\, dt \quad (\alpha > 0)$$

exists p.p. and is integrable.

(The function $f_\alpha(x)$ is the αth integral of $f(x)$. The definition agrees with the ordinary usage when α is a positive integer, and is valid when α is not an integer.)

(3) If $f_{\alpha,\beta}(x)$ is the βth integral of $f_\alpha(x)$, prove that

$$f_{\alpha,\beta}(x) = f_{\alpha+\beta}(x), \quad (\alpha > 0, \beta > 0)$$

if the right-hand side exists.

(4) If f, g are integrable, the function

$$r(x) = \int f(x-y)\, g(y)\, dy$$

is called the *resultant* of f and g (sometimes *convolution*, German *Faltung*).

Taking two cases,

(i) f and g periodic, and the integrals taken over a period,

(ii) integrals over $(-\infty, \infty)$,

prove that $r(x)$ is integrable, and that

$$\int |r(x)|\, dx \leqslant \int |f(x)|\, dx \int |g(x)|\, dx.$$

(5) If $f_n \to f(L^p)$, then $\int |f_n|^p \to \int |f|^p$.

(6) If $F(x)$ is the integral of a function of L^p, then, as $h \to 0$,

$$F(x+h) - F(x) = o(|h|^{1/p'}).$$

(7) Prove that, if f is in L^p for all p, then, as $p \to \infty$,

$$N_p(f) \to \max.|f|,$$

where $\max.|f|$ is the 'essential upper bound' of $|f|$, i.e. the smallest number M for which $|f| \leqslant M$ p.p.

This gives an interpretation of $p = 1$, $p' = \infty$ as conjugates, where $L^{p'}$ is the class of functions which are bounded (except in a set of measure zero).

CHAPTER VI

THE LEBESGUE-STIELTJES INTEGRAL

6·1. Integration with respect to a function. The idea of replacing the variable of integration x in an integral $\int f(x)\,dx$ by a function $\phi(x)$ of the variable is due to Stieltjes (1894). The simplest definition of a Stieltjes integral $\int_a^b f(x)\,d\phi(x)$ is as the limit of approximative sums

(1) $$\Sigma f(\xi_r)\{\phi(x_r) - \phi(x_{r-1})\},$$

where the x_r are points of subdivision of (a, b) and $x_{r-1} \leqslant \xi_r \leqslant x_r$. The existence of the integral can be proved if suitable assumptions are made about the functions f, ϕ, of which the most natural are that $f(x)$ is continuous and $\phi(x)$ monotonic (or of bounded variation).

As an illustration let $\phi(x)$ be a function which is discontinuous at isolated points $x = x_n$, with $\phi(x_n + 0) - \phi(x_n - 0) = k_n$, and whose graph consists otherwise of horizontal stretches. It is easy to see that the value of the integral $\int f\,d\phi$, taken over an interval containing the points x_n, is $\Sigma k_n f(x_n)$. One of the advantages of the Stieltjes integral is that it thus includes as special cases series of discrete terms. It is often convenient to use a physical mode of speech. An increasing function $\phi(x)$ determines a 'distribution of mass' along the x-axis, which can include continuous distributions and also masses concentrated at certain points (see Ch. IV, Ex. 6 for the detailed analysis of $\phi(x)$).

In this chapter we shall set up a 'Lebesgue-Stieltjes' integral. Our first task is to develop a theory of *variation of $\phi(x)$ over sets of points*. This will generalize the theory of measure and will reduce to it for the particular function $\phi(x) = x$.

6·2. The variation of an increasing function. Let $\phi(x)$ be an increasing function. It is now desirable to have a notation which distinguishes the open interval $a < x < b$ from the closed interval $a \leqslant x \leqslant b$; we shall write (a, b) for the former and $[a, b]$ for the latter.

If I is the open interval (a, b), we define the variation of ϕ over I, written $v(\phi, I)$ or $v(\phi, (a, b))$ or simply vI if it is clear what function ϕ is being considered,

$$vI = \phi(b - 0) - \phi(a + 0).$$

If O is an open set, vO is defined to be ΣvI summed over the intervals I of O.

With this beginning, a theory of variation can be set up on the model of Chapter II. In the first place, variation over a closed set is defined 'by complements'. We observe that, for a closed interval,

$$v(\phi, [a, b]) = \phi(b + 0) - \phi(a - 0),$$

and if the interval reduces to a single point x,

$$v(\phi, x) = \phi(x + 0) - \phi(x - 0)$$

which vanishes if and only if ϕ is continuous at x.

Following § 2·5, we define v^*E as the lower bound of vO for O's containing E, and v_*E as the upper bound of vQ for Q's contained in E. If $v^*E = v_*E$, it is convenient to say that E is *measurable* (ϕ), and ϕ then has a definite variation vE over E. The additive property holds for an enumerable infinity of sets $E_1, E_2, \ldots$, and also the analogous properties set out in § 2·8, § 2·9.

We shall not deal with variation of functions of more than one variable (§ 2·10). There is no difficulty of principle, but the discussion of a general plane mass-distribution, containing as it may masses concentrated at points and along curves involves complications of detail.

The concept of measurability (B) (see p. 23) is here of some importance. In general, a measurable set may or may not be measurable for a particular ϕ. A set which is measurable (B) is however measurable for every ϕ. For open and closed sets are measurable (ϕ), and so are sets obtainable from them by elementary operations including that of taking a limit (§ 2·9).

A function $f(x)$ is said to be measurable (ϕ) if the set of values of x for which $f(x) > A$ is measurable (ϕ) for every A. From the last paragraph, a function which is measurable (B) is measurable with respect to every increasing ϕ.

6·3. The Lebesgue-Stieltjes integral.

We assume ϕ increasing, and, for the present, f positive. It is useful to have both a geometrical representation of the integral by means of ordinate-sets and an analytical expression as the limit of approximative sums. We start with the former.

Let $\xi = \phi(x)$. Then to an interval (a, b) of Ox corresponds an interval (α, β) of $O\xi$. The correspondence between x and ξ is one-one, except that

(i) if $\phi(x)$ is constant in an interval (c, d), the same value of ξ corresponds to every value of x in (c, d),

(ii) if $\phi(x)$ has a discontinuity at x_0, we agree that every value of ξ between $\phi(x_0 - 0)$ and $\phi(x_0 + 0)$ shall correspond to x_0.

The function $x = \phi^{-1}(\xi)$, or $x = \chi(\xi)$ say, inverse to $\xi = \phi(x)$ is uniquely defined except for the values of ξ specified in (i) which form at most an enumerable set. Throughout this chapter we shall reserve the notation $x = \chi(\xi)$ for the inverse of $\xi = \phi(x)$.

To a given function $f(x)$ corresponds a function of ξ, $f\{\chi(\xi)\}$. If we wish to make the definition of this function unique we may agree that, for values of ξ of type (i), f shall be equal to the lower bound of the aggregate of values of the corresponding $f(x)$; for purposes of integration, however, indeterminacy in an enumerable set will not matter. We observe that a discontinuity of $\phi(x)$ corresponds to an interval of constancy of $\chi(\xi)$ and so of $f\{\chi(\xi)\}$.

We can now adopt as the definition of the Lebesgue-Stieltjes (LS) integral $\int_a^b f d\phi$ the value of the Lebesgue integral $\int_\alpha^\beta f\{\chi(\xi)\} d\xi$, that is to say, the measure of the ordinate-set of f having as its base the set of values of ξ, i.e. of $\phi(x)$.

In the same way $\int_E f d\phi$ is defined to mean $\int_{\mathscr{E}} f\{\chi(\xi)\} d\xi$ where $\mathscr{E}$ is the set of ξ corresponding to the set E of x.

If f takes both positive and negative values and $f = f_+ - f_-$, as in § 3·1, then $\int f d\phi$ is defined to be $\int f_+ d\phi - \int f_- d\phi$.

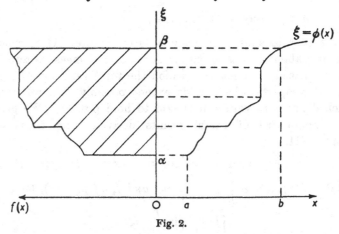

Fig. 2.

Text-fig. 2 illustrates (in the first quadrant) an increasing function $\xi = \phi(x)$ having a horizontal stretch and a discontinuity. The graph of it is 'projected' on to the ξ axis, and ordinates of a function $f(x)$ (shown as positive and increasing) are set up on the projection. The shaded area represents $\int_a^b f d\phi$.

F

The alternative method of definition of the integral of a bounded f is as the common limit of approximative sums

$$S = \sum_1^n l_{r+1} e_r, \quad s = \sum_1^n l_r e_r,$$

where e_r is the variation of ϕ over the set E_r for which $l_r \leqslant f(x) < l_{r+1}$ (cf. § 3·4). Extension to unbounded f is made as in § 3·5.

If ϕ is a function of bounded variation, then $\phi = \phi_1 - \phi_2$ where ϕ_1 and ϕ_2 are increasing functions (they are taken to be the positive and negative variations of ϕ). We then define

$$\int f d\phi = \int f d\phi_1 - \int f d\phi_2.$$

If the integrals on the right exist, so does $\int f d\phi_1 + \int f d\phi_2$ and this is appropriately written $\int f | d\phi |$.

It is easy to adapt the arguments of Chapter III to yield theorems about the LS integral. A set of x of measure zero is to be replaced by a set over which the total variation of ϕ is zero. A property which holds except possibly in a set over which ϕ has zero variation is said to hold p.p. (ϕ). We give two illustrations of useful results derived from those of Chapter III.

THEOREM OF DOMINATED CONVERGENCE. *If, for all* n, $|f_n(x)| \leqslant \psi(x)$, *where* $\int \psi | d\phi |$ *exists, and* $f_n \to f$ *p.p.* (ϕ), *then*

$$\int f d\phi = \lim \int f_n d\phi.$$

The theorem of bounded convergence is a special case of this.

INTEGRATION OF SERIES. *If either*

$$\int (\Sigma | u_n |) | d\phi | \quad \text{or} \quad \Sigma \int | u_n | | d\phi |$$

is finite, then

$$\int (\Sigma u_n) d\phi = \Sigma \int u_n d\phi.$$

Examples

(1) Establish the equivalence of the 'geometrical' and the 'approximative sum' definitions of the *LS* integral.

(2) Prove that

$$\int f d\phi_1 + \int f d\phi_2 = \int f d(\phi_1 + \phi_2).$$

(3) (The first mean-value theorem.) If ϕ is increasing, and $A \leqslant f \leqslant B$, then

$$A v(\phi, E) \leqslant \int_E f d\phi \leqslant B v(\phi, E).$$

(4) If $|f| \leqslant M$, then $\left| \int_E f d\phi \right| \leqslant M \int_E |d\phi|.$

6·4. Integration by parts. This theorem for Stieltjes integrals takes the elegant form

$$(1) \qquad \int_a^b f d\phi + \int_a^b \phi df = f(b)\phi(b) - f(a)\phi(a).$$

We shall investigate under what hypotheses on f and ϕ this holds.

If the integrals are defined in the original sense of Stieltjes (§ 6·1 (1)), then, if either integral on the left-hand side exists, so does the other and the formula is true.

Suppose that $\int f d\phi$ exists.

Let $a = x_0 \leqslant \xi_1 \leqslant x_1 \leqslant \ldots \leqslant x_{n-1} \leqslant \xi_n \leqslant x_n = b$ be any dissection of (a, b). Define $\xi_0 = a$, $\xi_{n+1} = b$.

Then if

$$T = \sum_1^n \phi(\xi_r)\{f(x_r) - f(x_{r-1})\},$$

and

$$T_1 = \sum_1^{n+1} f(x_{r-1})\{\phi(\xi_r) - \phi(\xi_{r-1})\},$$

we have identically,

$$T + T_1 = f(b)\phi(b) - f(a)\phi(a).$$

Observe that if either $\max(x_r - x_{r-1})$ or $\max(\xi_r - \xi_{r-1})$ tends to 0, so does the other.

Since $\int_a^b f d\phi$ exists, it is the limit of T_1 as $\max(x_r - x_{r-1})$ tends to 0. Therefore T tends to a limit, and so $\int \phi\, df$ exists and relation (1) holds.

We now link this up with the LS integral.

If f, ϕ have bounded variation and there is no value of x for which they are both discontinuous, then, if a, b are points of continuity of f, ϕ, the formula (1) holds.

(If a, b are discontinuities of f or ϕ, the right-hand side of (1) is to be replaced by

$$f(b-0)\,\phi(b-0) - f(a+0)\,\phi(a+0),$$

or $$f(b+0)\,\phi(b+0) - f(a-0)\,\phi(a-0),$$

according as the interval is open or closed.)

The general case can be built up from monotonic functions f, ϕ, and it will be sufficient to give the detail for increasing f, ϕ.

Since f, ϕ have no common discontinuities, given ϵ, we can take a subdivision

$$a = x_0 < x_1 \ldots x_{n-1} < x_n = b$$

so fine that in any (x_{r-1}, x_r) at least one of $f(x_r) - f(x_{r-1})$ and $\phi(x_r) - \phi(x_{r-1})$ is less than ϵ.

The Stieltjes sum $\Sigma f(\xi_r)\{\phi(x_r) - \phi(x_{r-1})\}$ lies between

$$U = \Sigma f(x_r)\{\phi(x_r) - \phi(x_{r-1})\}$$

and $$u = \Sigma f(x_{r-1})\{\phi(x_r) - \phi(x_{r-1})\},$$

where $$U - u < \epsilon\{\phi(b) - \phi(a) + f(b) - f(a)\}.$$

Moreover U decreases and u increases (in the wide sense) when further points of subdivision are inserted. So by the argument of § 3·4, U and u have a common limit as $\max(x_r - x_{r-1})$ tends to 0, and the Stieltjes integral exists.

There is a pair of approximative sums for the LS integral $\int f \, d\phi$ lying between U and u, and the result follows.

The theorem has a simple geometrical interpretation. The reader should draw a diagram taking, say, f and ϕ both to be positive and increasing. Then, with values of f and ϕ measured along two perpendicular axes, the ordinate-sets defining the integrals fill up the part of the plane between the two rectangles whose opposite vertices are the origin and the points $\{f(a), \phi(a)\}$ and $\{f(b), \phi(b)\}$ respectively.

Example

Develop this geometrical interpretation into a proof of the theorem.

6·5. Change of variable. Second mean-value theorem.

If $\phi(x)$ is of bounded variation and $\Phi(x) = \int_a^x g \, d\phi$, then

$$\int_a^b f \, d\Phi = \int_a^b f g \, d\phi.$$

We may suppose that ϕ is increasing and that f and g are of constant sign (say both positive).

Write $\xi = \phi(x)$ and $\Xi = \Phi(x)$, and let their inverse functions be

$$x = \chi(\xi) \quad \text{and} \quad x = X(\Xi)$$

respectively.

Then, by definitions of the integrals as ordinate-sets, we have

$$\Xi = \int g\{\chi(\xi)\} \, d\xi,$$

$$\int f \, d\Phi = \int f\{X(\Xi)\} \, d\Xi,$$

$$\int f g \, d\phi = \int f\{\chi(\xi)\} g\{\chi(\xi)\} \, d\xi,$$

the integrals being taken between the appropriate limits.

The result will now follow from the theorem of change of variable for Lebesgue integrals (§ 5·2), if we satisfy ourselves that the integrals are unaffected by the many-valuedness of the inverse functions χ and X. We find that g may be undefined in an enumerable set of ξ, and f in an enumerable set of Ξ, corresponding to intervals of constancy of $\phi(x)$; f may also be undefined in an interval in which $g = 0$ (so that Ξ is constant); both the integrals vanish over such an interval.

We now prove the second mean-value theorem (for the L integral).

If f is monotonic and g integrable, then

$$\int_a^b fg\,dx = f(a)\int_a^\xi g\,dx + f(b)\int_\xi^b g\,dx$$

for some ξ satisfying $a \leqslant \xi \leqslant b$, where a, b are values for which f is continuous.

(If a, b are discontinuities of f, $f(a)$ and $f(b)$ on the right-hand side are to be replaced by $f(a+0)$ and $f(b-0)$.)

Let $$G(x) = \int_a^x g\,dx.$$

Then the LS integral $\int f dG$ exists and we have by the last theorem,

$$\int_a^b fg\,dx = \int_a^b f dG$$

$$= \left[fG\right]_a^b - \int_a^b G\,df, \quad \text{by integration by parts,}$$

$$= f(b)\,G(b) - G(\xi)\{f(b) - f(a)\},$$

by the first mean-value theorem (§ 6·3, Ex. 3), and this is equal to

$$f(a)\int_a^\xi g\,dx + f(b)\int_\xi^b g\,dx.$$

The brief account that we have given of Stieltjes integrals should enable the reader to manipulate them with confidence.

Differential properties (depending on the notion of differentiating with respect to a function ϕ) do not often come into question.

EXAMPLES ON CHAPTER VI

(1) If $\phi_n \to \phi$, state sufficient conditions for

$$\int_a^b f\,d\phi_n \to \int_a^b f\,d\phi.$$

(2) Construct a Stieltjes-Fubini theorem (§ 5·5).

(3) Prove that

$$\int_0^1 \frac{d\phi(x)}{x} = \sum_{n=m}^\infty \binom{n}{m}\int_0^1 x^m(1-x)^{n-m}\,d\phi(x).$$

SOLUTIONS OF SOME EXAMPLES

Hints are given for the solutions of all but the easiest examples, and more detailed solutions of those which are most important.

CHAPTER I

P. 9, Ex. 6. Let E_n be the set of x such that $f(\xi) < f(x)$ for $x - 1/n < \xi < x$ and $x < \xi < x + 1/n$. Each E_n is isolated and ΣE_n is the set of maxima of $f(x)$.

P. 9, Ex. 8. Expressed as infinite decimals in the scale of 3, the points of E are represented by

$$\cdot a_1 a_2 a_3 \ldots a_n \ldots,$$

where the a's are either 0 or 2 (not 1). There is a one-one correspondence between these numbers and

$$\cdot b_1 b_2 b_3 \ldots b_n \ldots$$

in the scale of 2, defined by $b_n = 0$ or 1 according as $a_n = 0$ or 2. These include all the numbers of $(0, 1)$.

CHAPTER II

P. 25, Ex. 1. The fraction of the interval $(0, 1)$ left after the nth operation is $(2/3)^n$.

P. 25, Ex. 2. Consider rational values of y.

P. 25, Ex. 3. We construct first a *step-function* $\psi(x)$ (i.e. ψ is constant in each of a finite number of interval which together make up (a, b)), which satisfies the inequalities postulated for ϕ.

Take K so large that the set E_K in which $|f| \geqslant K$ has measure $< \frac{1}{2}\epsilon$. Divide the range $(-K, K)$ into n parts each of length $< \epsilon$. Let (l_{r-1}, l_r) be a typical part. Let E_r be the set of x for which $l_{r-1} \leqslant f < l_r$. E_r is measurable. Let $E_r = \mathscr{E}_r + e_r' - e_r''$, where $\mathscr{E}_r$ is a finite set of intervals and $me_r' < \epsilon/4n$, $me_r'' < \epsilon/4n$.

Define $\psi_r = l_r$ in $\mathcal{E}_r$ and 0 elsewhere. Then $\psi = \sum_1^n \psi_r$ is a step-function which differs from f by less than ϵ except in $E_K + \sum_1^n (e'_r + e''_r)$ and this set has measure less than ϵ.

Any step-function can be modified into a continuous function differing from it in a set of arbitrarily small measure. For, if ξ is a typical point of discontinuity of ψ, with $\psi(\xi - 0) = k$ and $\psi(\xi + 0) = l$, define ϕ in the interval $(\xi - \eta, \xi + \eta)$—where η is arbitrarily small—as the linear segment joining $(\xi - \eta, k)$ to $(\xi + \eta, l)$; outside intervals $(\xi - \eta, \xi + \eta)$, keep $\phi = \psi$.

Further conditions could be imposed on ϕ, e.g. the possession of a derivative of any order; we could satisfy them by 'rounding off the corners' of the ϕ already constructed.

P. 25, Ex. 4. Write $g_n = |f - f_n|$. Let ϵ_r be a decreasing sequence tending to 0. Let $E_{n,r}$ be the subset of E in which $g_\nu < \epsilon_r$ for $\nu \geqslant n$. Then $E_{n,r} \subset E_{n+1,r}$ and $\lim_{n \to \infty} E_{n,r} = E$ (since, given r, every point of E belongs to some $E_{n,r}$). Hence, given r, we can find $n(r)$ such that

$$m(E - E_{n(r),r}) < \frac{\delta}{2^r}.$$

Write $E_0 = E_{n(1),1} E_{n(2),2} \dots$. In E_0, for all values of r, $g_n < \epsilon_r$ for $n \geqslant n(r)$, i.e. $g_n \to 0$ uniformly. And

$$m(E - E_0) \leqslant \sum_{r=1}^{\infty} m(E - E_{n(r),r}).$$

CHAPTER III

§ 3·5, p. 33, Ex. 3.

$$f(x) = 2x \sin \frac{1}{x^2} - \frac{2}{x} \cos \frac{1}{x^2} \quad (x \neq 0),$$

$$f(0) = 0.$$

The first term of $f(x)$ is continuous and integrable. For the second term, let I_n be the interval of x defined by

$$2n\pi - \frac{\pi}{3} < \frac{1}{x^2} < 2n\pi + \frac{\pi}{3}.$$

Its length $> \dfrac{A}{n\sqrt{n}}$. In I_n, $\cos\dfrac{1}{x^2} \geqslant \dfrac{1}{2}$ and $\displaystyle\int_{I_n}\left|\dfrac{2}{x}\cos\dfrac{1}{x^2}\right|dx > \dfrac{A}{n}$. If $\displaystyle\int_0^1\left|\dfrac{2}{x}\cos\dfrac{1}{x^2}\right|dx$ existed, it would be $> \displaystyle\sum^\infty\int_{I_n}$.

§ 3·7, p. 37. (1) g bounded *or* (2) f^2, g^2 integrable. If (1), then fg is measurable and, if $|g| \leqslant K$, integrability of fg follows from that of $K|f|$. If (2), use $2|fg| \leqslant f^2 + g^2$.

P. 42, Ex. 2. $f_n(x) = nxe^{-nx^2}$ in $(0,1)$.

P. 42, Ex. 3. Let H be a subset, with $mH > mE - \delta$, in which $f_n \to f$ uniformly. Then there is $n_0(\delta)$ such that $|f - f_n| < \delta$ in H for $n > n_0$.

$$\int_E |f - f_n| = \int_H + \int_{E-H},$$

$$\left|\int_{E-H}\right| \leqslant \{\text{upper bound of } |f - f_n|\}\, m(E - H) < 2M\delta,$$

$$\left|\int_H\right| \leqslant \delta mE \text{ for } n > n_0.$$

Hence $\qquad\qquad \displaystyle\int |f - f_n| \to 0$ as $n \to \infty$.

P. 42, Ex. 4. $\left(1 - \dfrac{x}{n}\right)^n < e^{-x}$ and $\to e^{-x}$. Use theorem of dominated convergence. (Observe how the use of the Lebesgue integral avoids the tiresome inequality work needed to discuss uniformity of convergence.)

P. 43, Ex. 9. Take an enumerable sequence $\Sigma^{(1)}, \dots, \Sigma^{(n)}, \dots$ of subdivisions of (a,b), the points of $\Sigma^{(n)}$ containing those of $\Sigma^{(n-1)}$. Let $\delta_r^{(n)}$ be an interval of $\Sigma^{(n)}$, $M_r^{(n)}$ the upper bound of $f(x)$ in $\delta_r^{(n)}$. Let $\lim_{n\to\infty}\max \delta_r^{(n)} = 0$.

Define $\Phi_n(x) = M_r^{(n)}$ in $\delta_r^{(n)}$ for all r of $\Sigma^{(n)}$.

If E is the (enumerable) set of all points of division in $\Sigma^{(1)}, \dots, \Sigma^{(n)}, \dots$, then, except in E,

$$\Phi_n(x) \to M(x).$$

$\Phi_n(x)$ is measurable. Let $n \to \infty$ and apply the theorem of bounded convergence.

P. 43, Ex. 10. $M(x) = m(x)$ p.p.
P. 43, Ex. 11. See Ch. II, Ex. 3.
P. 43, Ex. 12. Use Ex. 11.

CHAPTER IV

P. 57, Ex. 1. Let $f(x)$, $g(x)$ be absolutely continuous in (a,b) and $\Sigma(x_r, x_r + h_r)$ a set of disjoint intervals in (a,b).

$$\Sigma \,|\, f(x_r + h_r)\, g(x_r + h_r) - f(x_r)\, g(x_r) \,|$$
$$= \Sigma \,|\, f(x_r + h_r)\, \{g(x_r + h_r) - g(x_r)\} + g(x_r)\, \{f(x_r + h_r) - f(x_r)\} \,|$$
$$\leqslant M \Sigma \,|\, g(x_r + h_r) - g(x_r) \,| + N \Sigma \,|\, f(x_r + h_r) - f(x_r) \,|,$$

where M, N are the upper bounds of $|f|$, $|g|$ in (a,b). These last two sums tend to zero with Σh_r.

P. 57, Ex. 2. $1 = f(1) - f(0) = \Sigma_1 \Delta f + \Sigma_2 \Delta f$, where Σ_1 is taken over the intervals of lengths $\dfrac{1}{3}, \dfrac{1}{3^2}, \ldots, \dfrac{1}{3^k}$ specified in the construction and Σ_2 over the intervals complementary to them. By definition $\Sigma_1 = 0$ and so $\Sigma_2 = 1$. But the intervals of Σ_2 have total length $(2/3)^k$ and this tends to 0 as $k \to \infty$.

P. 57, Ex. 3. *First proof* (using Vitali's theorem). Suppose that there is a sub-set E_1 of E with $mE_1 > 0$ in which E has not density 1. Then there is an $\alpha < 1$, such that the set E_2 at points of which $\varliminf \dfrac{m(EH)}{h} < \alpha$, where H is one of the intervals $(x-h, x)$, $(x, x+h)$, has $mE_2 = e_2$ (say) > 0.

Given ϵ, enclose E_2 in O, $mO < e_2 + \epsilon$.

By Vitali's theorem, there is a set $\mathscr{E}(= I_1 + \ldots + I_n)$ of disjoint intervals with $m\mathscr{E} > e_2 - \epsilon$ in each of which $m(I_r E) < \alpha m I_r$.

The sub-set of E_2 in $\mathscr{E}$ has measure $< \alpha m \mathscr{E}$; the sub-set of E_2 in $O - \mathscr{E}$ has measure $< 2\epsilon$.

Hence $e_2 < \alpha(e_2 + \epsilon) + 2\epsilon$, which is false if ϵ is small enough.

Second proof. This, due to Lebesgue, starts from scratch and contains its own 'covering' argument.

LEMMA. *Let O be an open set. The set of left-hand points of intervals in which O has average density greater than $1/K$ has measure KmO, where $K > 1$.*

Let $(a_1, b_1), \ldots, (a_n, b_n), \ldots$ be the intervals of O. Extend the interval (a_1, b_1) to the left by taking (α_1, b_1) such that $b_1 - \alpha_1 = K(b_1 - a_1)$. Do the same to (a_2, b_2), giving (α_2, b_2). If $(\alpha_1, b_1), (\alpha_2, b_2)$ have an interval in common, move this to the left so as to give an interval (λ, μ) containing (α_1, b_1) and (α_2, b_2) of length equal to $(b_1 - \alpha_1) + (b_2 - \alpha_2)$. We have thus either one or two intervals, say $(A_1, B_1), (A_2, B_2)$.

Carry out the same construction with (a_3, b_3) giving (α_3, b_3). If (α_3, b_3) has an interval in common with either (A_1, B_1) or (A_2, B_2), carry it to the left as before. We then have either one, two or three intervals.

This construction leads to an enumerable set Ω of non-overlapping intervals $(\lambda_1, \mu_1), \ldots, (\lambda_n, \mu_n), \ldots$ of total measure KmO. The set Ω is such that, for any (λ, μ) inside (λ_r, μ_r), the measure of the set of points of (λ, μ) not in O is at least $(K-1)$ times the measure of the set common to O and (λ, μ). Therefore, in an interval (λ, μ) of which the left-hand end-point does not belong to Ω, the average density of O is at at most $1/K$.

To prove the density theorem, we first show that an open set O has zero density p.p. in CO. Take p intervals of O, say O^p, with $mO^p > mO - \epsilon$. Given K, construct the set Ω of the lemma for the set $O - O_p$. Then

$$m(O_p + \Omega) \leqslant mO_p + Km(O - O_p)$$

$$\leqslant mO + (K-1)\epsilon.$$

All the points of CO at which the right-hand density of O is greater than $1/K$ lie in a set of measure at most $(K-1)\epsilon$. Since ϵ is arbitrary they form a set of measure zero. Let $K \to \infty$; the right-hand density of O is zero p.p. in CO. Similarly so is the left-hand density.

Let now E be any measurable set; we prove that its density is zero p.p. in CE. Enclose E in O with $m(O - E) < \epsilon$. The set of points of CO in which E has not zero density is contained in the set in which O has not zero density, i.e. has measure zero. Since ϵ is arbitrary, E has density zero p.p. in CE.

Interchange the roles of E and CE. CE has density zero p.p. in E and so E has density 1 p.p. in E.

P. 57, Ex. 4. Let $|f| < K$. Let $E_{u,v} = $ set where $u \leqslant f < v$. Let $N_{u,v}$ be the set of points of $E_{u,v}$ which are not points of density. Then $N = \Sigma N_{u,v}$ summed over all rationals u, v with $u < v$ has measure zero. We prove that $F' = f$ for x in CN.

Let x be a point of $E_{u,v}$ and I an interval containing x. Then

$$-Km(I \cdot CE_{u,v}) + um(I \cdot E_{u,v}) \leqslant \int_I f$$
$$\leqslant Km(I \cdot CE_{u,v}) + vm(I \cdot E_{u,v}).$$

Divide by mI and let $mI \to 0$. Using the density theorem we have $u \leqslant \overline{\lim} \dfrac{1}{mI} \displaystyle\int_I f \leqslant v$.

P. 57, Ex. 5. Let $|\phi'(x)| < K$. Then

$$\left| \frac{\phi(x+h) - \phi(x)}{h} \right| = |\phi'(x + \theta h)| < K.$$

Let h take a sequence of values tending to 0. $\dfrac{\phi(x+h) - \phi(x)}{h} \to \phi'(x)$ boundedly. Argue as in § 4·7.

P. 57, Ex. 6. Prove $f_1 = f - \chi$ is (1) continuous, (2) increasing. Then $\phi = \displaystyle\int f_1'$.

CHAPTER V

P. 68, Ex. 1. (i) Repeated integrals unequal, (ii) double integral exists if $\alpha < 2$, (iii) $\int dy \int dx$ exists; $\int dx \int dy$ does not.

P. 69, Ex. 2.

$$\int_0^a f_\alpha(x)\,dx = \frac{1}{\Gamma(\alpha)} \int_0^a dx \int_0^x (x-t)^{\alpha-1} f(t)\,dt.$$

By Fubini, this

$$= \frac{1}{\Gamma(\alpha)} \int_0^a dt \int_t^a (x-t)^{\alpha-1} f(t) \, dx$$

$$= \frac{1}{\Gamma(\alpha)} \int_0^a f(t) \frac{(a-t)^\alpha}{\alpha} \, dt = \frac{1}{\Gamma(\alpha+1)} f_{\alpha+1}(a),$$

since the latter repeated integral exists when the integrand is replaced by its modulus.

P. 69, Ex. 3.

$$f_{\alpha,\beta}(x) = \frac{1}{\Gamma(\alpha)\,\Gamma(\beta)} \int_0^x (x-t)^{\beta-1} \, dt \int_0^t (t-u)^{\alpha-1} f(u) \, du$$

$$= \frac{1}{\Gamma(\alpha)\Gamma(\beta)} \int_0^x f(u) \, du \int_u^x (x-t)^{\beta-1} (t-u)^{\alpha-1} \, dt \quad \text{(justify)}.$$

In the inner integral put $t-u = v(x-u)$ and it becomes

$$(x-u)^{\alpha+\beta-1} \int_0^1 (1-v)^{\beta-1} v^{\alpha-1} \, dv = (x-u)^{\alpha+\beta-1} \frac{\Gamma(\alpha)\,\Gamma(\beta)}{\Gamma(\alpha+\beta)}.$$

P. 69, Ex. 4.

$$\int r(x) \, dx = \int g(y) \, dy \int f(x-y) \, dx = \int g(y) \, dy \int f(x) \, dx,$$

$$\int |r(x)| \, dx \leqslant \int |g(y)| \, dy \int |f(x)| \, dx,$$

the inversions of the repeated integrals being justified by Fubini.

P. 69, Ex. 5. $N_p(f) - N_p(f-f_n) \leqslant N_p(f_n) \leqslant N_p(f) + N_p(f-f_n).$

P. 69, Ex. 6. Apply Hölder to $\displaystyle\int_x^{x+h} f(t) \, dt.$

P. 69, Ex. 7. $N_p(f) \leqslant M$ is clear. $|f| > M - \epsilon$ in E_1 where $mE_1 > \delta$, and so $N_p(f) \geqslant \delta^{1/p} (M - \epsilon)$ and this $\to M - \epsilon$ as $p \to \infty$.

CHAPTER VI

§ 6·4, p. 77. Consider the ordinate-set defining $\displaystyle\int_a^b f \, d\phi$ (diagram of § 6·3); their tops are $\xi = \phi(x), \eta = f(x)$. The function

$\eta = f\{\chi(\xi)\}$ is a continuous function of ξ except (1) for ξ corresponding to an interval of constancy of ϕ, or (2) for $\xi = \phi(x)$, where $f(x)$ is discontinuous.

For each ξ of type (1) or (2) there is a linear segment defined by $\eta(\xi - 0) \leqslant \eta \leqslant \eta(\xi + 0)$.

Adjoining these segments to the set of points forming tops of ordinates, we have a continuous curve (in general with horizontal and vertical stretches). This same curve is obtained from the tops of ordinates of $\int_a^b \phi\, df$.

P. 79, Ex. 1. f continuous, total variation of $(\phi - \phi_n)$ in (a, b) tends to 0 as $n \to \infty$ *or* obtain a set of conditions by integrating by parts.

P. 79, Ex. 2. $\int d\phi(x) \int f(x, y)\, d\psi(y) = \int d\psi(y) \int f(x, y)\, d\phi(x)$, provided that one of

$$\int |d\phi(x)| \int |f(x, y)|\, |d\psi(y)|$$

$$\int |d\psi(y)| \int |f(x, y)|\, |d\phi(x)|$$

is finite. From § 5·4.

P. 79, Ex. 3. The terms of $\sum_{n=m}^{\infty} \binom{n}{m} x^m (1 - x)^{n-m}$ are positive in $(0, 1)$ and the sum is $1/x$.

References for §2·7

J. E. LITTLEWOOD, Elements of the Theory of Real Functions (Dover Publications, 1954).

J. B. ROSSER, Logic for Mathematicians (McGraw-Hill, 1953).

W. SIERPIŃSKI, Leçons sur les nombres transfinis (Gauthier-Villars, 1928).

Printed in the United States
By Bookmasters